RAPTURE OF THE DEEP

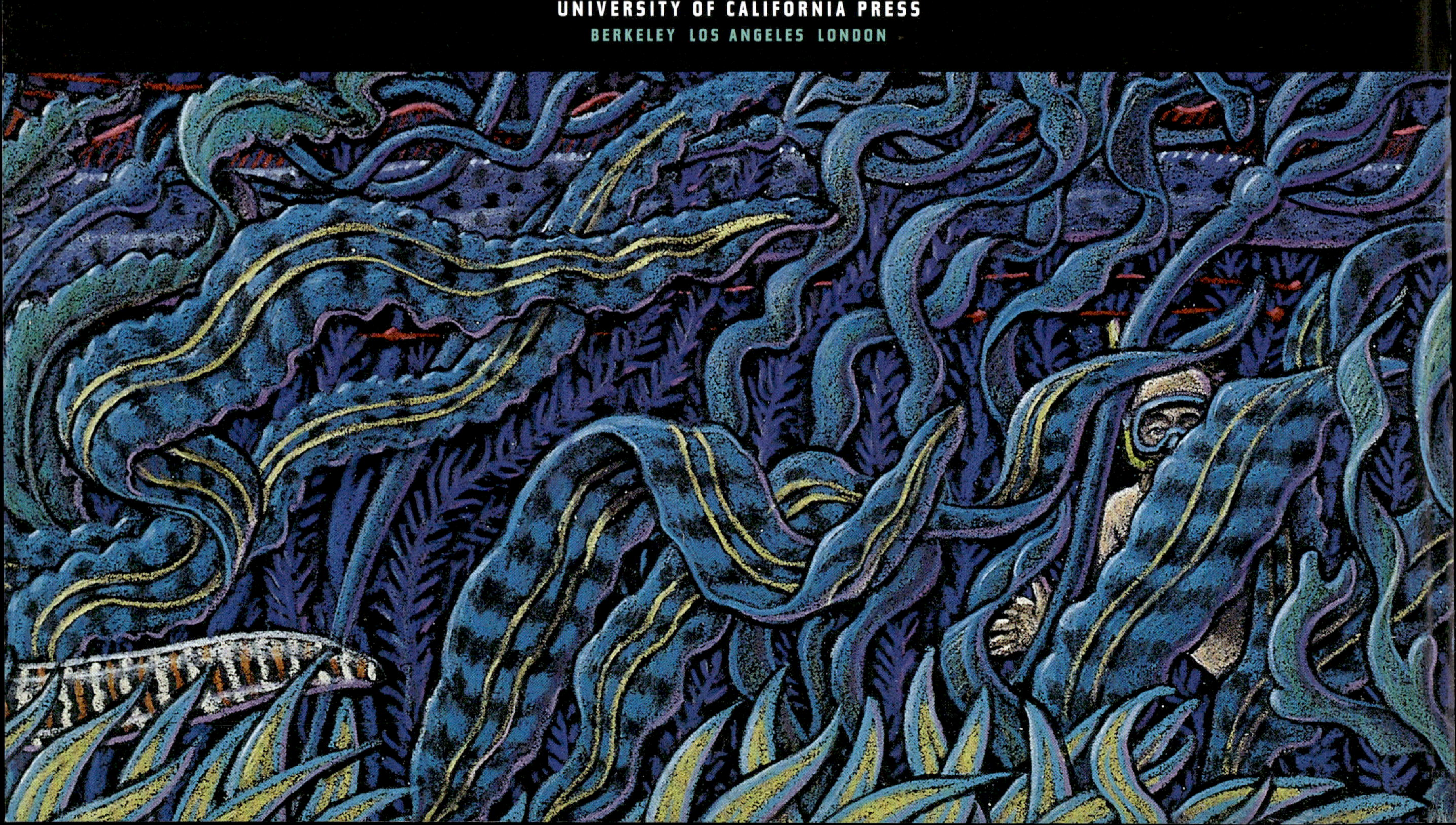
UNIVERSITY OF CALIFORNIA PRESS
BERKELEY LOS ANGELES LONDON

RAPTURE OF THE DEEP

THE ART OF RAY TROLL

WITH AN INTRODUCTION BY BRAD MATSEN AND COMMENTARY BY THE ARTIST · FOREWORD BY DAVID JAMES DUNCAN

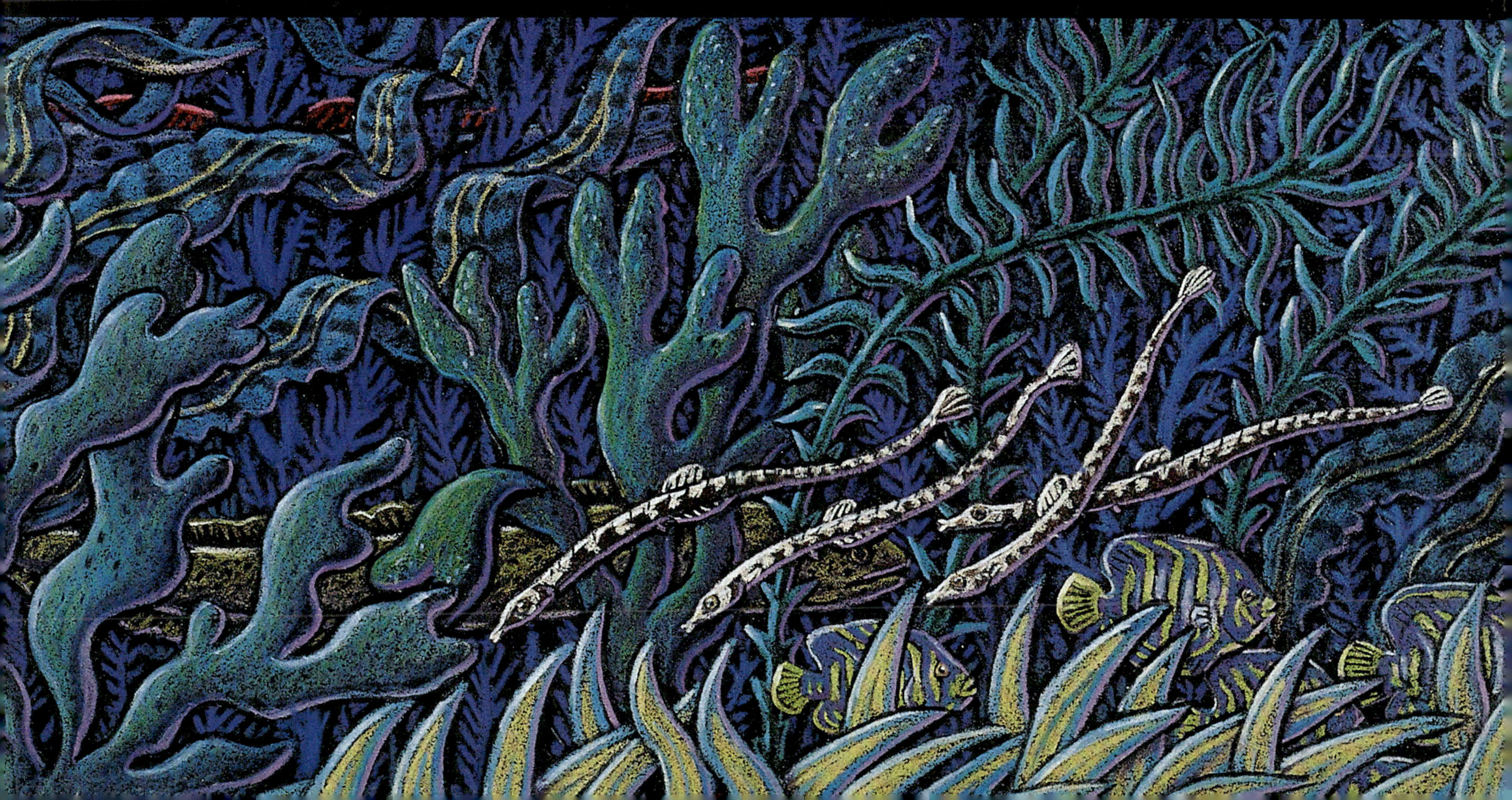

University of California Press
Berkeley and Los Angeles, California

University of California Press, Ltd.
London, England

Title page: Detail from *Kelp Is on the Way*; contents page: detail from *Deep Time Diver.* Both 2003, colored pencil on paper, 6 in. × 22 in., private collection; photos by Jim Wolon.

Library of Congress Cataloging-in-Publication Data

Troll, Ray, 1954–.
Rapture of the deep : the art of Ray Troll / with an introduction by Brad Matsen and commentary by the artist ; foreword by David James Duncan.
p. cm.
ISBN 0-520-23947-4 (cloth : alk. paper).
1. Troll, Ray, 1954—Catalogs. 2. Fishes in art—Catalogs.
I. Title: Art of Ray Troll. II. Title.

N6537.T69A4 2004
760'.92—dc22 2004004208

Manufactured in Canada
14 13 12 11 10 09 08 07 06 05
10 9 8 7 6 5 4 3 2 1

The paper used in this publication meets the minimum requirements of ANSI/NISO Z39.48-1992 (R 1997) (*Permanence of Paper*).

FOR MY BROTHER, TIM,
AND MY SISTERS, KATE, MIMI, TERRI, AND SUSIE

THE ARTIST AND THE SCIENTIST BRING OUT OF THE DARK VOID,
LIKE THE MYSTERIOUS UNIVERSE ITSELF, THE UNIQUE, THE STRANGE, THE UNEXPECTED.

Loren Eiseley, *The Mind as Nature*

ART IS ONLY A WAY TO LEARN HOW TO LIVE.

Henry Miller, *The Colossus of Maroussi*

CONTENTS

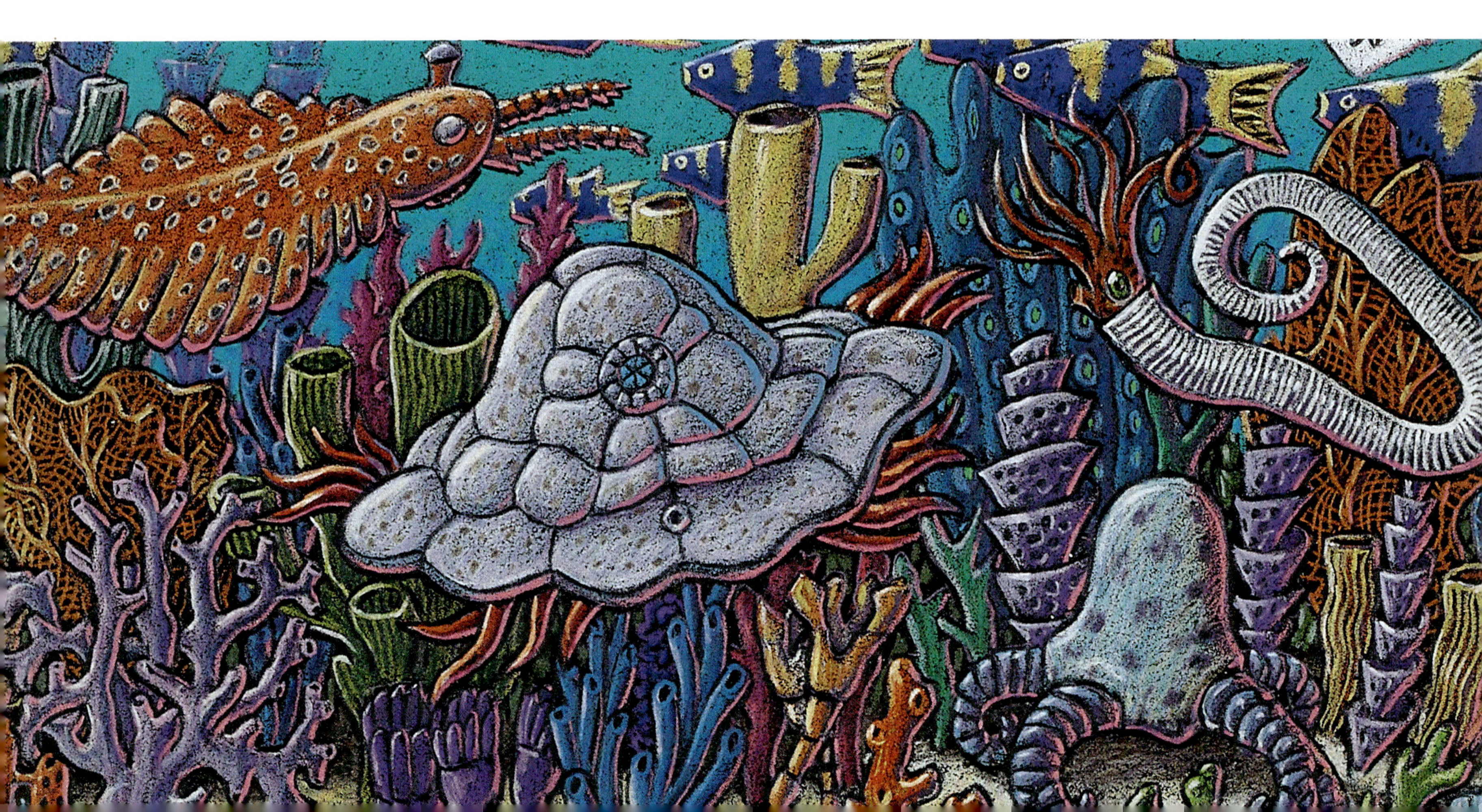

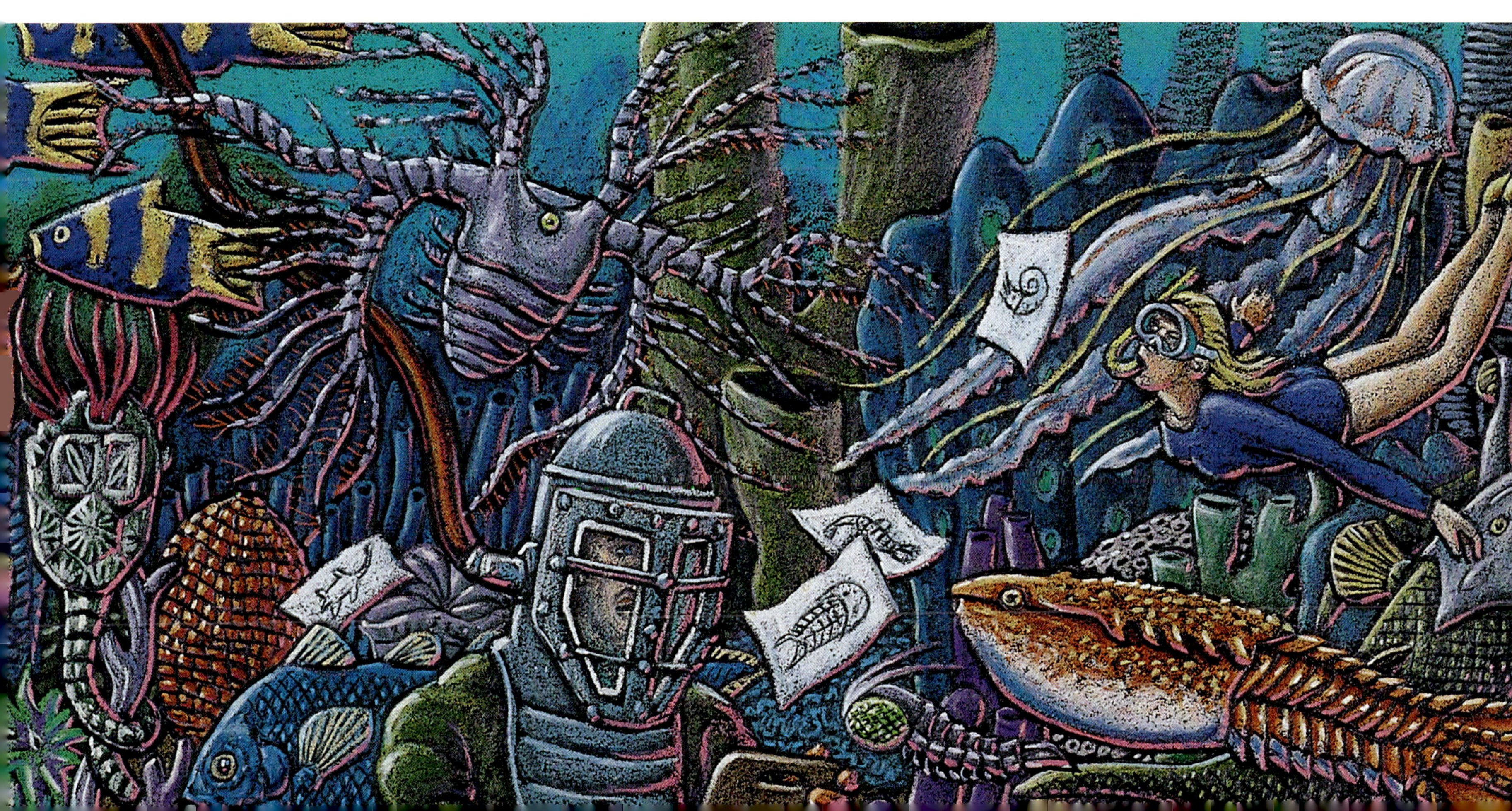

FOREWORD

THE NEO-FLEMISH DARWINIST OF METAPHYSAESTHETICS

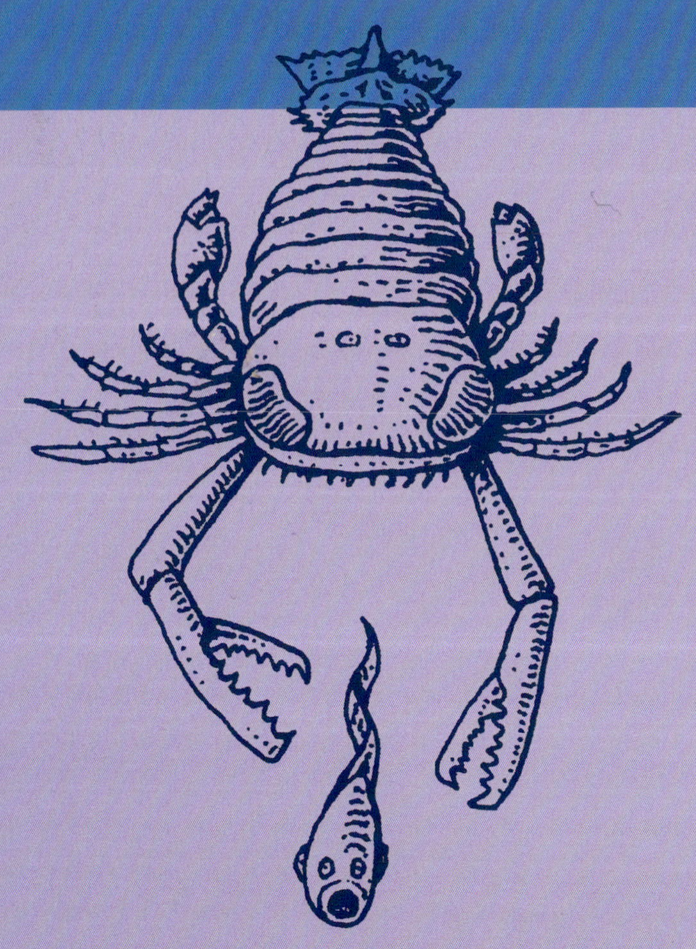

DAVID JAMES DUNCAN

RAY "RATFISH" TROLL

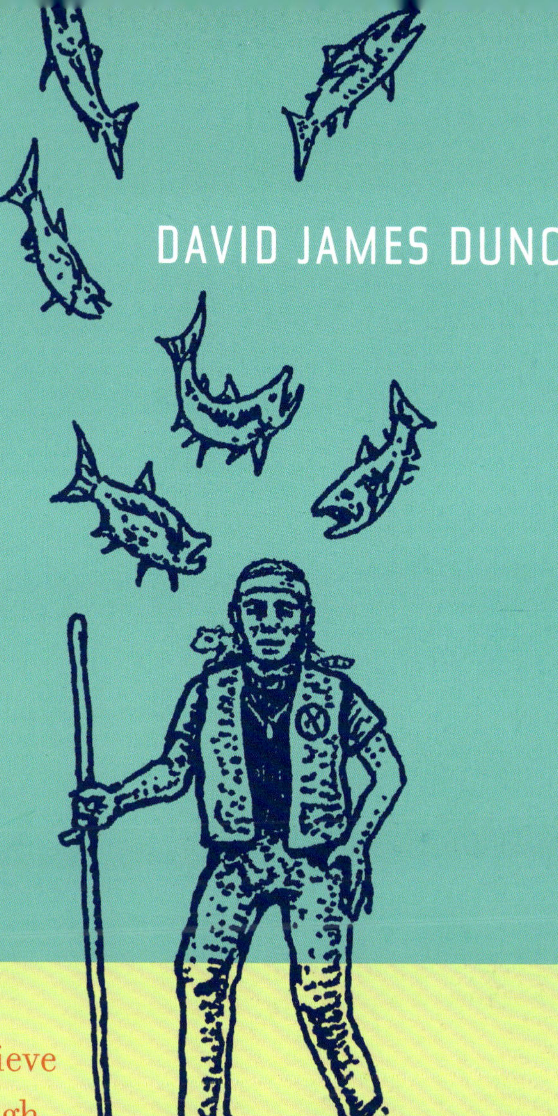

Sorry about that title—I can't pronounce it, either. I believe I'm on to a little something regarding Ray and his art, though. Allow me to begin by unpacking what I can't pronounce.

By "Ratfish" I refer to one of the most unfairly named, otherworldly, eerily beautiful species of fish I've ever dredged off an ocean floor. If a picture is worth a thousand words, a caught fish is worth ten thousand. The first ratfish I pulled up out of Puget Sound, at age ten, did not require the power of faerie or of speech in order to distinctly inform me: *I am an alien and magical creature from a world unknown to you. I need to eat in my world, so you have caught me. Now you must gently let me go. My galaxy has nothing to do with yours.* It says ten thousand words about Ray Troll that he uses this Outer Spaceling's name to nick his own.

By "Metaphysaesthetics" I mean what an art critic means by "aesthetics," with an added twist. Just as Ray, by nature, obsesses on paleontological ideas, I by nature obsess on metaphysical ideas. It is only natural, therefore, that my comments on Ray's aesthetics be a bit metaphysaesthetical.

By "Neo-Flemish" I refer to the likes of Hieronymus Bosch and Pieter Bruegel the Elder—two Flemish artists seen by the art critics of their day as too fanciful, lewd, or antic to be taken "seriously," yet so loved by nonacademicians and fellow artists that their fame became global. As has Ray's. It should also be noted that the very art critics who lamented Bruegel's and Bosch's lack of seriousness happened, in their spare time, to be bona fide Inquisitors fond of putting to death those whose religious and artistic visions veered too far from their own. What better way to outwit the fascist fanatics than to hide seriousness of intent behind an antic tone? By "Neo-Flemish" then, I refer to, say, the tone of a contemporary Troll-shirt, which, thanks to its jocular design and coloration, may be worn in public among the Neo-Inquisitors of our day despite its flagrant reference to erotic sex. Or a Troll poster of nihilistic gloom, made publicly presentable, even cheery, via the prophylactically obscure Latin tag, "NULLUM GRATUITUM PRANDIUM."

Finally, by "Darwinist" I refer to Charles. Darwin is so huge an influence on Ray that I want to expound briefly on the former, as a way of sneaking up on the art of the latter.

First of all, Darwin was a scientist, not a religious thinker. Although he is depicted by some as an antireligious deicide, the only "religion" his theories have damaged is a cultish fundamentalism well suited to languish in a world where the unfit do not survive. Born and raised a Christian, Darwin has nothing unkind to say about his Maker. True, his theories of natural selection and evolution soon smashed through Bilious Piety and Scientific Rote-Think like a wrecking ball. But he refrains, in his famed books, from all comment on religion.

There is, however, an overlooked fact about Darwin: the philosophers and religious scholars who actually studied the man, rather than condemn him to hell, found so much to like that they began playing with his theories in a manner I'm tempted to call "Troll-like." A Darwin contemporary and religious scholar named James Freeman Clarke, for instance, pointed out that Darwin "took no notice of the evolution of the soul, but only of the body." Because of this obsession with physical form, writes Clarke, Darwin never addresses the metaphysical possibility of an ongoing evolution of *consciousness* that may well occur simultaneously with the evolution of forms. Clarke was among the first to declare that evolutionary theory as propounded by Darwin could be powerfully wed to the Hindu/Buddhist theory of an undying nexus of consciousness, or soul, that is "developed and educated by passing through many bodies."

On another bad day for fundamentalists, Harvard philosopher Francis Bowen expanded on this riff, arguing that life on earth, "if limited to the duration of a single mortal body," was

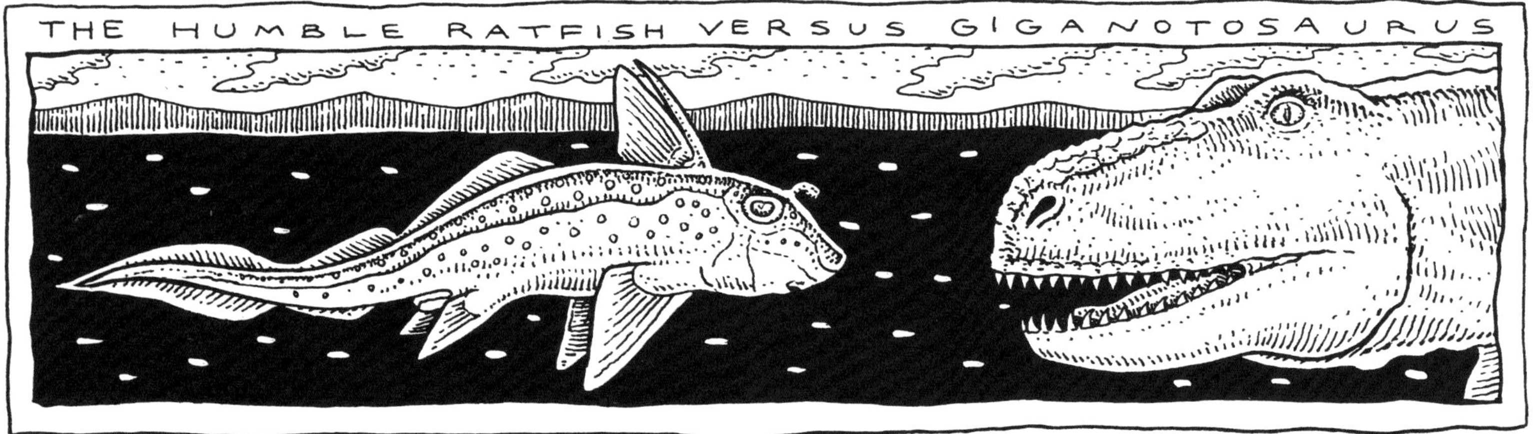

WHAT MAKES RATFISH SO COOL? BESIDES LOOKING LIKE A SCIENCE EXPERIMENT GONE WRONG, THESE FISH EVOLVED OVER 300 MILLION YEARS AGO. THAT'S WAY BEFORE THE DINOSAURS ROAMED THE EARTH. THE BIG DINOS ARE GONE, BUT RATFISH STILL SWIM IN THE OCEANS TODAY. RATFISH HAVE CHANGED VERY LITTLE OVER TIME SO THEY TRULY ARE LIVING FOSSILS. RATFISH ARE ALSO KNOWN AS CHIMAERAS. THEY ARE CARTILAGINOUS FISH AND ARE CLOSELY RELATED TO SHARKS. THERE ARE ABOUT FOURTY SPECIES ALIVE TODAY.

surely "an inadequate preparation for eternity." And what proof do we have, Bowen asked, that the sojourn of our souls is confined to such narrow limits? Why couldn't the soul-journey, as he put it, "be repeated through a long series of successive generations, the same personality animating one after another of an indefinite number of tenements of flesh, and carrying forward into each the training it has received [and] the character it has formed?"

Many Darwin-era authors echo this refrain. In his journals, Thoreau writes of a part of himself that, "as it were, is not a part of me, but a spectator, sharing no experience, but taking note of it. . . . [And] when the play, it may be the tragedy, of life is over, the spectator goes his way." In a letter to a friend, Louisa May Alcott opines that "immortality is the passing of a soul through many lives or experiences; and such as are truly lived, used, and learned, help on to the next." A century after Darwin published his theory, Boris Pasternak's protagonist in *Doctor Zhivago* says to a child afraid to fall asleep, "You are anxious about whether you will rise from the dead or not, but you rose from the dead when you were born and didn't even notice. There is nothing to fear. There is no such thing as death." As John Muir wrote, with characteristically rhapsodic glee, "This grand show is eternal!"

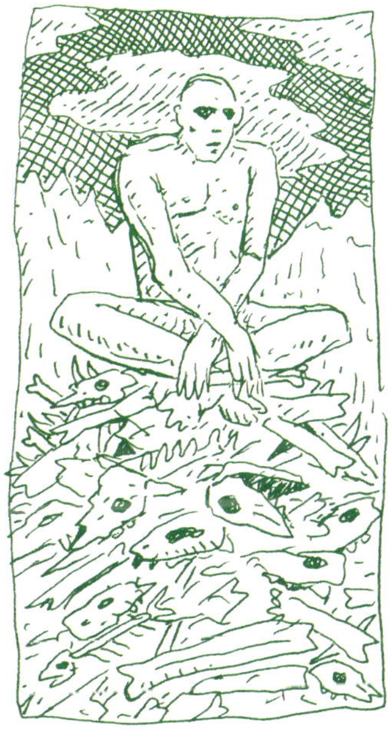

Darwin did not appear to see this kind of speculation coming, but it's fascinating when an individual's work ends up having repercussions different from any they imagined. Darwin's effort to demonstrate a few truths regarding biological organisms exploded itsy-bitsy intellectual pieties and opened doors by the thousand—and once doors have been opened, who knows what might walk through next? Darwinian evolution will never mesh with the fundamentalist concept of a single brief life leading to a Judgment Day. But it meshes so gracefully with the evolution of *consciousness* implied by Eastern concepts of reincarnation and karma that it could be argued that Darwin—however unintentionally—helped open the West to the spiritual riches of the East.

DOMINION

To mention such matters in the context of an art book by and about Ratfish Ray may set some eyes rolling—Ray's included. But like Bruegel, Bosch, and Darwin, Ray is an exploder of religious balderdash and an opener of doors—and no more in control of what then steps through those doors than Darwin. I've only met Ray once, but there are three metaphysaesthetical facts I know about him via means so mysterious I can't say how I know them, yet I know them for sure. Here they are:

(1) If Ray Troll were John Muir, he would have cried out, "This grand show of fishes is eternal!"

(2) If Ray had penned *Doctor Zhivago*, his protagonist would have consoled the sleepless child by saying, "There is nothing to fear. You used to be a fish."

(3) In Ray's oeuvre, the species we are likeliest to continue to see animating one "tenement" of art after another, carrying forward into each the colors, forms, traits, training, and character this species and Ray have symbiotically formed, are fish. Which seems weird at first. But when a single fascination is so obsessively, creatively, oft-hilariously portrayed, certain rarified pleasures begin to accrue. For me, these pleasures happen to be Neo-Flemish, Darwinist, and (sorry Ray!) Metaphysaesthetical.

Ray is a philosopher in the visual, as opposed to the verbal, tradition of artists like Bruegel, who feel compelled to say little more than "it's a Fish-Eat-Fish World out there" before they shut up, hunker down, and start depicting how such a world looks and feels. The last time I was in New York, though, there

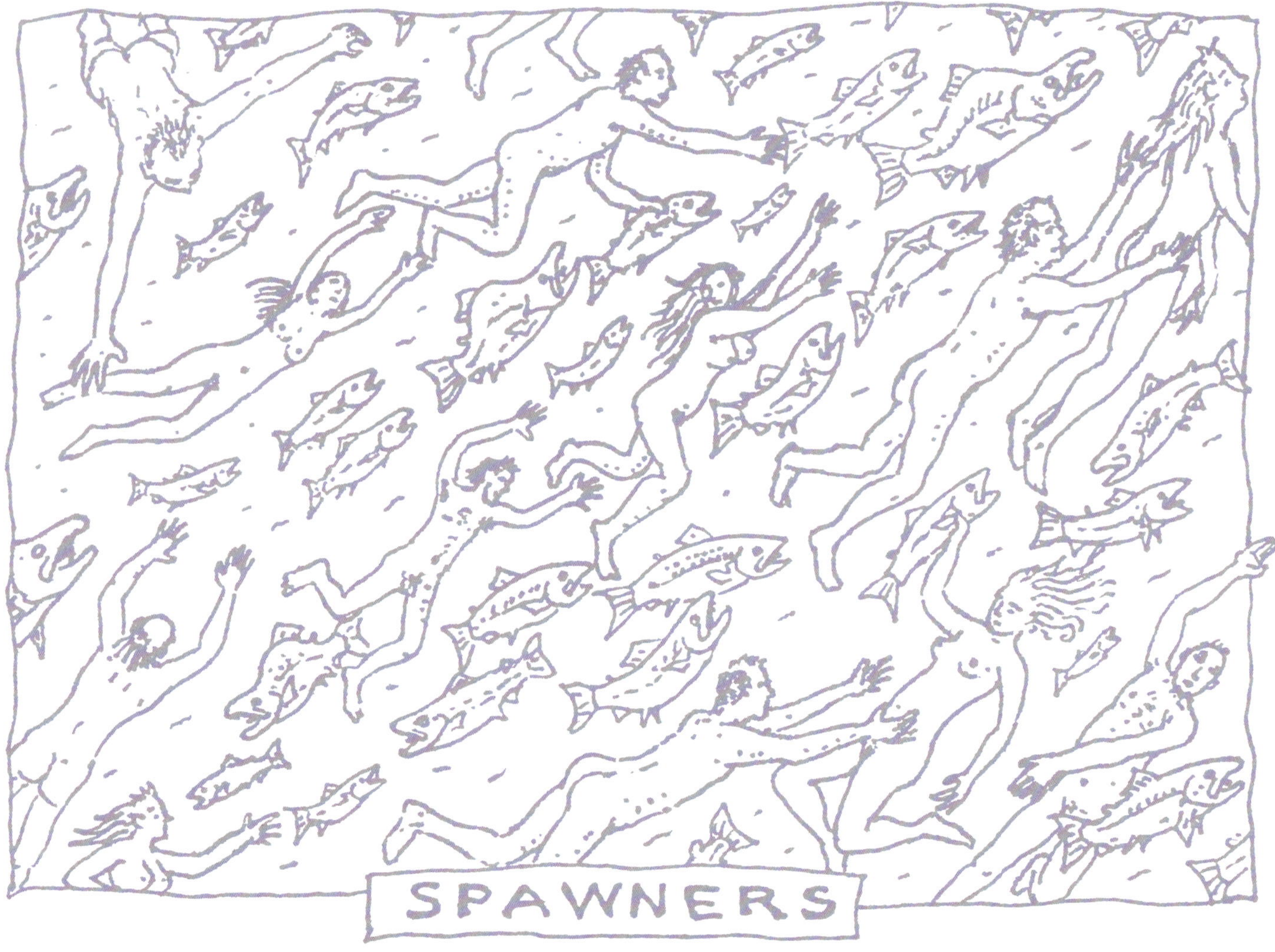

was a huge Bruegel show at the Metropolitan Museum, and after spending a couple of hours among the socket-eyed possum-faces, impossibly bent spines, hybrid reptile/rat/demon/human bodies, rectum-shaped mouths, mouth-shaped rectums, endlessly vacant expressions, and endlessly ruthless follies, I realized that for all his amazing talent, Bruegel's vision of humanity makes me nauseous.

Ray's vision of humanity does nothing of the sort. Ray lays claim to no highfalutin *or* lowfalutin reasons for his obsession with fish. He simply set out years ago, in a disciplined artistic fashion, to demonstrate a few truths, metaphors, spectacular colors and patterns, remarkable forms, visual ironies, and shameless puns inseparable from these creatures. As a responsible husband and father he chose to make a living doing it, hence the posters and T-shirts. And as a responsible rock and roller his target was never some feckin' art critic: it was

Ketchikan Everyman. But Fishes Past, Present, Future, Freshwater and Salt, as Ray so meticulously and happily depicts them, reveal a wealth of forms so much more colorful, varied, and charming than a strict version of natural selection or a Bruegelian notion of life would lead us to expect, that even the way Ray's fish eat his other fish tends to yield a feeling of delight.

Bruegel portrays fish and people devouring fish because people are fishlike and odious. Troll portrays fish and people devouring fish because people are fishlike and *marvelous*. Big difference! In Ray's work, the artist is gleeful at sharing a world with, and resembling, fish. In Ray's mind, the thought of incarnating as a ratfish is so cool that he named himself after this remote possibility. In Ray's art as it is on earth, and who the heck knows about heaven, the greatest of salmon species—the Chinook, a.k.a. *King*—shares its name with crabs, cards, flowers, shells, snakes, trees, creeks, cartoon characters, a state-of-the-art tank, a tennis player, a plethora of birds and fishes, an even greater plethora of human royalty, extinct reptiles, America's greatest Civil Rights hero, a checkerboard move, a fast-food chain, and Elvis the Pelvis, and it's all good. In Ray's metaphysaesthetic, the recognition that "it's a Fish-Eat-Fish World out there" is not depressing; it's *consoling*, for it means, among other things, that *we get to eat lots of yummy fishes!*

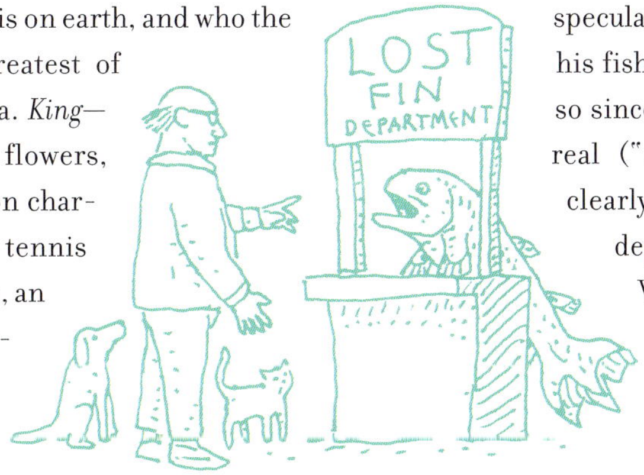

When I study the way Ray depicts fish, the fish in my own life become more vivid and delicious and dear to me. When I ponder the length and detail of his artistic engagement with fish, and then consider the resilience and complexity of the creatures capable of inspiring such an engagement, I feel far fewer penultimate, and *zero* ultimate, fears for the intergalactic Future of Fishes, Life, Ray, me, and Fish-lovers everywhere.

Ray is as blithely unconcerned with metaphysical speculation as an artist of his caliber can be, yet his fish are so ancient and stupendously varied, so sincere ("Spawn Till You Die"), so ruthlessly real ("NULLUM GRATUITUM PRANDIUM"), so clearly seen, and above all so loved by their depicter that they somehow "show us The Way."

Which is a metaphysaesthetical way of saying: *Hey! Thanks, Ratfish Ray!*

INTRODUCTION

A PORTRAIT OF THE ARTIST

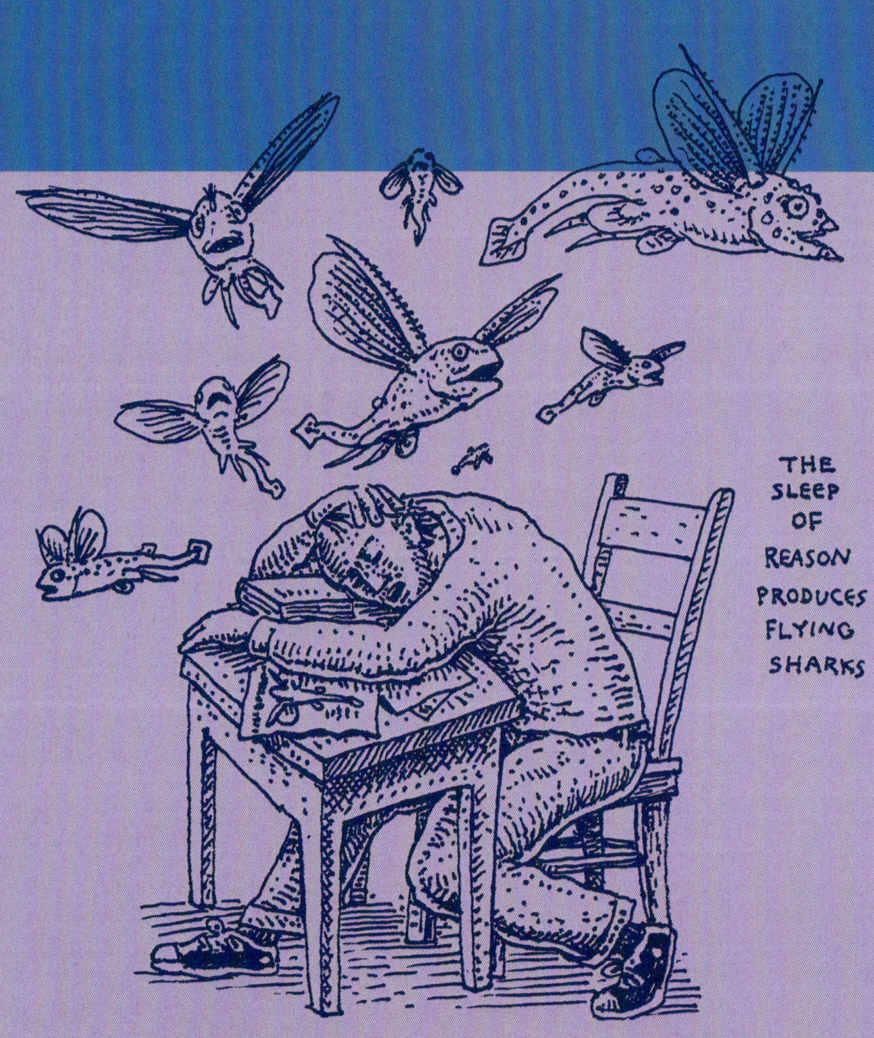

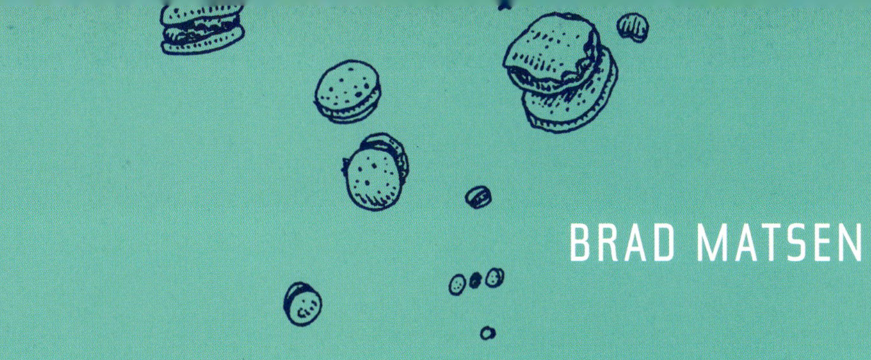

BRAD MATSEN

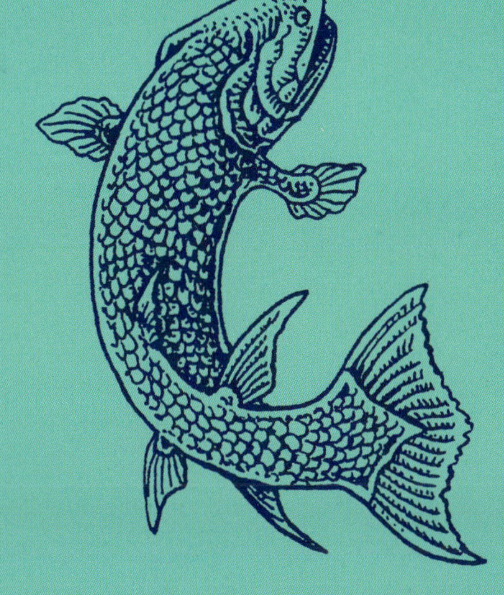

ON THE ROAD

Ray Troll and I left Seattle on our first road trip of the twenty-first century in a rented Taurus with a cooler full of Frontier Room barbecue, about fifty CDs, and the wildly optimistic plan of driving six hundred miles to the motel next to Lulu's Bar and Grill in Redding, California, in one day. Since it was already ten a.m., we would pass over the Siskiyous south of Ashland and down onto the Mt. Shasta plateau just after dark, with a full moon rising that January night over the eerie white pyramid of the volcano. Then we'd slide through the Shasta lakes, which would be shimmering like chrome in the moonlight, and from there it would be only sixty miles to Redding, food, and sleep. The three hundred miles from Redding to Monterey the next day would be a breeze; we'd have three days there, a couple of days doing business in Berkeley, then two for the trip back to Seattle. It was a good plan.

The deal was that whoever wasn't driving got to pick the music. We played our traditional road trip anthem, Springsteen's "Born to Run," as we whipped through the Mercer Street tunnel and up onto the interstate; then Ray scanned through my CDs, put them back in the glove box, and never looked at them again. I didn't mind, though, because if I didn't hang out with him once in a while I'd only be listening to stuff like Gordon Lightfoot and opera. Ray is an over-the-falls rock-and-roll fanatic. He has a monthly radio show on KRBD in Ketchikan, Alaska, where he lives, and likes nothing more than to be invited to act as the disc jockey at the town's legendary dance parties. Ray and his two kids play together in a band called Ray's Big Idea—or RBI—and the first time I laid eyes on him about twenty years ago he was thumping the guitar in a local band called the Squawking Fish. He emails an annual "Top Five CDs" list to his hundreds of friends and has a "Top Five Desert Island CDs" list. And of course, he consumes music while he works. When he looks at one of his paintings or drawings he can tell you what he was listening to when he made it. His stupendous Amazon mural (which he still owns), for instance, is Lucinda Williams, Steve Earle, and a Brazilian band called Os Mutantes (The Mutants).

Ray is a pretty regular guy when he's working. At eight o'clock every weekday morning he walks from the house—where he's just finished breakfast with his wife, Michelle, and his children, Corinna and Patrick—to his three-story, brick-red studio next door. He loads his five-disc CD player and, with the discipline of a medieval stone carver, paints, sketches, and draws until precisely noon, when he walks back to the house for a sandwich. At one he returns to the studio and stays there until four. As one of our outdoorsy friends says, Ray would be very easy to trap because his routines are so predictable.

He is anything but predictable in his musical tastes, though. In the car on I-5 he thumbs through his own CD wallet and reads me names and titles like a professor briefing a student on the curriculum: the Flaming Lips; P. J. Harvey; Steve Earle; Lucinda Williams; Patty Griffin; REM; David Byrne; Pulp; Low; the Stones; R. L. Burnside; the Beatles' *Revolver* (which Ray says is the number-one rock-and-roll record of all time); Joseph Arthur; the Future Bible Heroes; the Magnetic Fields; *The Royal Tennenbaums* soundtrack, featuring Nico; and the Yeah, Yeah, Yeahs, Corinna's favorite new band.

The Life of Troll

Raymond Michael Troll was born on March 4, 1954 (a Pisces, of course), in Corning, New York, the third of six children in the Catholic family of Raymond and Mary Troll. His father was a B-29 navigator in World War II, got out of the Air Force, went to law school, and joined the Air Force again. The Trolls lived on air bases, including Tachikawa in Japan—what we then referred to as "the Far East"—during the waning years of the American occupation. Colonel Troll finally retired and moved his family to Kansas, where he worked for an airplane company during Ray's high school years.

Ray was an art kid all along. At eleven he was already drawing elaborate depictions of Cortez's capture of Tenochtitlan in 1521; the loss of the Spanish Armada in 1588; the Battle of Shiloh in 1862; and D-Day on Omaha Beach on June 6, 1944. (One of his most recent drawings is of the Battle of Old Sitka, a Tlingit attack on a Russian-Aleut fort in Southeast Alaska in 1802.) Ray also drew the usual boy stuff—spaceships, airplanes, and dinosaurs.

In another stranger-than-fiction foreshadowing, when he was eight Ray and his brother Tim used to get together with a few of their buddies at their grandparents' house to play "Museum." While all the other kids were out playing baseball or shooting dice or whatever they did in Corning, New York, in the early sixties, Troll and his pals were laying out their little collection of arrowheads, bones, and other local artifacts and charging their parents, brothers, and sisters a nickel to look at them. A photographer from the *Corning Leader* actually came over and took a picture of Ray and his museum, which the paper ran on the front page.

Ray was a teenager in Kansas when he met his first art mentor, Harold Caldwell, whose precise renditions of fossils and other critters kindled a fascination for scientific accuracy in the boy. Harold produced educational film strips, and Ray became his assistant. The idea that a person could make a few bucks drawing pictures apparently wasn't lost on Ray; he survived the rest of high school in conservative Kansas wrapped in the identity of the outsider artiste, and then went on to study art nearby in Lindsborg, at Bethany College. There he picked up a B.A. in printmaking, refining the trajectory of his life that almost twenty-five years later would put him in the passenger seat of a gold Taurus SSE on southbound I-5 with me.

We made good time out of Seattle thanks to cruise control and the seventy-mile-an-hour speed limit, and settled into the sweet reality of road slack, with Ray giving a running commentary on his CDs—the names of the musicians' former bands and other rock-and-roll arcana, which to me is a little like listening to somebody speak Hungarian. In no time we were through Portland and into the Willamette Valley, gnawing on the barbecue and watching northbound swarms of geese organizing themselves into those undulating vees that are so mesmerizing, even if you've seen them a thousand times. The Rousseau-green valley spreads into soggy wetlands to the west that are the main line for the annual migrations of the geese, swans, cranes, egrets, herons, and ducks. During the three hours it took us to drive from Salem to the Siskiyous there were few moments when we weren't looking at a sky of geese or fields transformed into pointillist extravaganzas of resting white herons and egrets. Though neither of us knows much about the timing of bird migrations, we were pretty sure they were going to end up in Alaska in a few months; we turned up the Stones singing "Faraway Eyes" ("Before Ronnie Woods they went through a couple of other lead guitarists," etc., etc.) and started thinking about the miracle of that place in our lives. A kind of churchy feeling settled over the cabin of the Taurus as we watched the northbound birds.

Art in Alaska

The great good fortune that sent Ray north to Alaska is as much a part of his creative past as his drawings of historic battles, his make-believe museum, or his printmaking degree. After he graduated from Bethany College, Ray moved to Seattle, mainly because it was a bigger city than Wichita and in the late seventies was the nexus of the hip track of the Pacific Coast art scene. There he got a dose of another art-world reality: the day job. He worked as a waiter at the Aurora Tavern, silk-screened bus placards with underprivileged kids, and for a while was an intake agent for the Internal Revenue Service, answering phones in an office building eight hours a day. In Seattle Ray also picked up on the stimulating urban blend of music, the counterculture in full bloom, and some pretty good artists taking their licks during the twilight of the American frontier. Deciding to stick around the Pacific Northwest, he applied for the Master of Fine Arts program at Washington State University, three hundred miles east over the mountains, in Pullman. He was twenty-six years old, as hungry for art as a wolf for a lamb chop, and showing promise as a printmaker, draftsman, and painter. He was admitted to the class of 1981, and off he went into the embrace of graduate school.

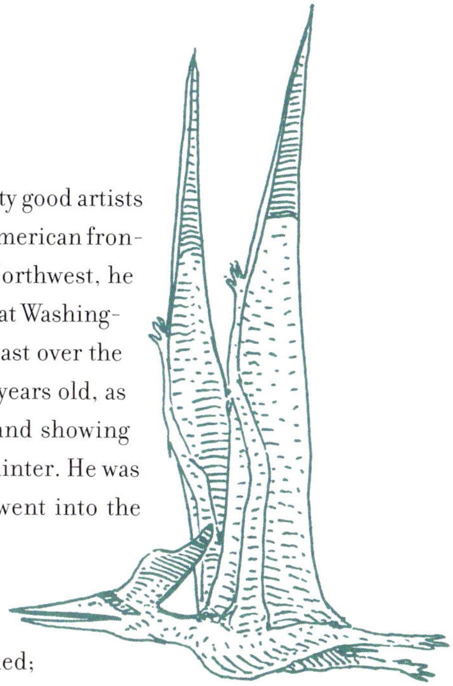

In Pullman Ray did what most MFA candidates do: he taught a little art to undergraduates; drew, painted, and photographed; attended lectures by Gaylen Hansen, Robert Helm, and the other faculty luminaries; and played in a band called Zuzu and the Robot Slave Boys, led by another artist, Jo Hockenhull. You don't really learn a hell of a lot in the formal sense in graduate art school because most of the time you're hanging around with other like-minded people doing what you'd be doing anyplace else. Almost everybody who has survived an MFA program, however, can recall some watershed event that propelled him or her out of art or further into it—either way a good outcome.

Ray caught one of his art breaks in Pullman when an art professor named Pat Siler devastated him with a C in his drawing class. "I fester on that grade to this day," Ray told me. "But when Pat Siler told me that my line lacked character and that my images were too easy, it pissed me off enough to make me focus. And I buckled down and got better." Getting a little miffed still works to stir Ray to inspired resolution, and he remembers that C in drawing as a turning point in his life as an artist.

Not only did Ray's drawing improve dramatically, but it became the central medium for much of his future work. Art critic Hilton Kramer called drawing "the chamber music of the visual arts." "Just as in a musical composition for a piano trio, say, or a string quartet," Kramer wrote, "we're better able to attend not only to the separate instruments but to the individual notes than in a symphonic work for full orchestra, so with a drawing on paper, our concentration is more sharply focused on the lines and other marks and touches of the artist's hand than on the drawing to be discerned in a large oil painting on canvas. Drawings place us in an intimate, almost conversational relation to the artist's sensibility."

Ray is most at home with pencils and pens. Most of what you'll see in this book was executed with pencils from the Prismacolor spectrum of 126 colors. He also works in charcoal, scratchboard, and ink, these days with Tombo pens or pigment markers that can take a wash. Often he works on black paper, progressing from the highlights down to give an effect he calls "black velvet cheesy." He uses scratchboard to much the same effect.

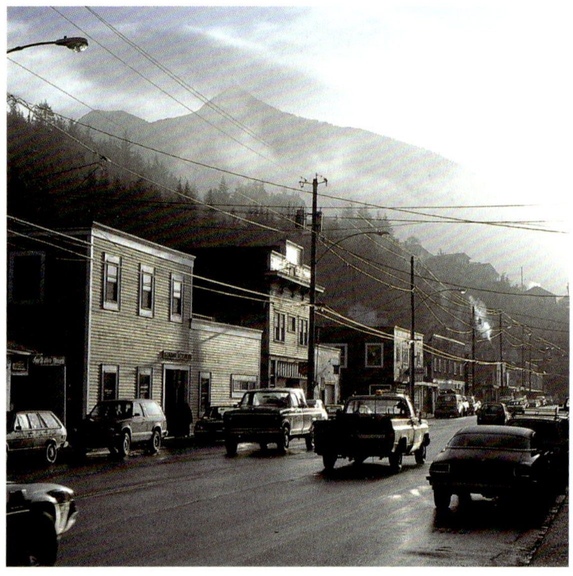

Ketchikan, a town of 14,000 people on Southeast Alaska's panhandle, has been Ray Troll's home for the last twenty years. This is a view of Water Street, where he had his studio in an old cannery building for ten of those years. That's Deer Mountain looming in the background.

Another thing a lot of people who finish an MFA program will tell you is that getting the degree can drop you into some very difficult territory. You've had your show, kissed all the girls, left town, and now what? For many, the prospects for making a living doing art are so daunting that teaching others to make art is an attractive alternative, however absurd it might seem (since your students, too, eventually discover that the prospects for making a living . . . etc., etc.). So Ray went back to Kansas to teach at his undergraduate alma mater, Bethany College. There he drank beer at the Öl Stuga, where they still remember him, and eventually reached the conclusion that he'd better move on.

The rest of the Troll family had radiated from Kansas soon after Ray went west to Seattle, and through one of those convoluted sets of luck and circumstance that lead people to Alaska, his sister Kate had ended up in a little town at the southern extreme of the Alexander Archipelago named Ketchikan. At the time, in the early eighties, it was a backwater of the natural resource empire nourished by seasonal work in the spruce, hemlock, and cedar forests or on the fishing grounds where salmon, halibut, crab, and shrimp were still in pretty good shape. Like the other towns of Southeast Alaska—Juneau, Petersburg, Wrangell, Sitka, Haines, and the rest—Ketchikan was a well-kept secret, with oil money flowing into roads, schools, a state ferry system, and other creature comforts that took the sting out of the long, wet, dark winters. The native people—Tlingits, Haidas, and Tsimshians—whose ancestors have been inhabiting the spooky, verdant coast since the last ice age, are still there, and the elegance and complexity of their culture adds a dimension of ancient richness to the place.

Ketchikan is on Revillagigedo Island (you can use the name as a sobriety test before you start your car), just minutes or hours away, depending on whether you go by boat or by float plane, from Misty Fjords, the outside coast, and endless access to *pax natura.* The town is famous for rain—two hundred inches a year. Some people can't stand it and flee, but others stay. Those who do stay adjust to wearing full rain gear a lot of the time, getting depressed in the winter, spending innumerable hours entertaining themselves and their friends, and going to Mexico or Hawaii whenever possible. In return, they get spectacular vistas when the weather's good, as well as unlimited fishing, hunting, kayaking, and camping. The fourteen thousand or so people who live in Ketchikan navigate within a slightly schizophrenic society spiced by the constant rumbling of passionate environmentalists, loggers, fishermen, fundamentalist Christians, unreconstructed counterculture refugees, and assorted end-of-the-roaders. These days, commercial fishing is limping, the last big timber mill has closed down, and the usual paybacks at the end of a resource extraction orgy are coming due. Another current downside is the cruise ship boom that began a few years after Ray Troll landed on his sister's doorstep in Ketchikan—though if you sell art, crafts, or T-shirts, the cruise ships are a plus.

Ray arrived in Ketchikan in the spring of 1983 to work in his sister's retail fish market, spent some time on the slime line at one of the packing plants, taught drawing at the community college, and started sniffing the art wind from a rented studio above a cannery. For most people, the term "Alaskan art" has referred to the often elegant but always conservative landscapes and animal paintings of awestruck white explorers, settlers, and visitors, and the handsome carvings, crafts, and constructions of native Alaskans. It's all earnest, representative, and not very difficult—beautiful, sure, and necessary, but none of it has any edge to it. Just about the time Ray started pitching fish in Ketchikan, though, a different sort of art scene was just starting to cast its scent into the breeze.

Ray already had a pantheon of art heroes, including contemporaries such as Carl Chew in Seattle, Leonard Koscianski in Baltimore, and California's William Wiley, whose paintings and assemblages were built on whimsy, humor, and often outrageous puns. Ray's work and artistic sensibility had been shaped, too, by *Mad* magazine and Turok, Son of Stone, comics; as well as Edward Gorey, William Blake, Magritte, Escher, Jan van Eyck, Gauguin, Karl Bodmer, George Catlin, Hieronymus Bosch, Ed "Big Daddy" Roth, Rousseau, Max Beckmann, and Charles Knight. He was particularly inspired by Pieter Bruegel the Elder, the Flemish painter known for his landscapes and lively scenes of ordinary people at work and play that give us

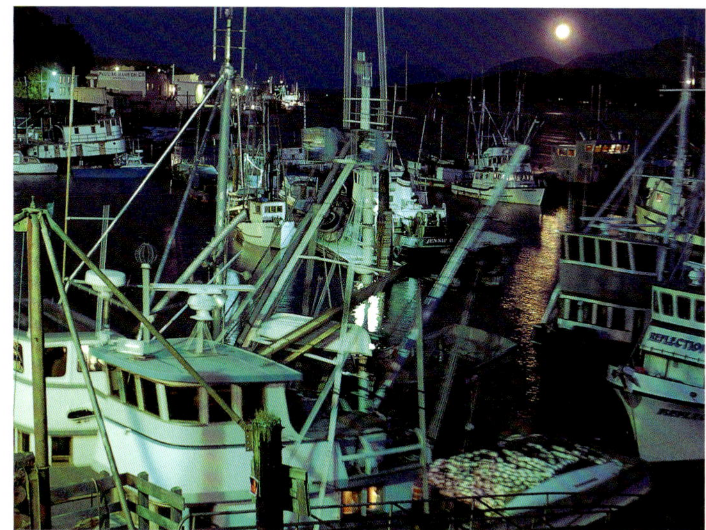

A view of the fishing fleet crowding the docks of Ketchikan on a rare clear midsummer night.

The "slime line" at a fish processing plant in Southeast Alaska. For a taste of what the scene is really all about, add a blaring rock-and-roll beat, wear plastic pants, and wet yourself down.

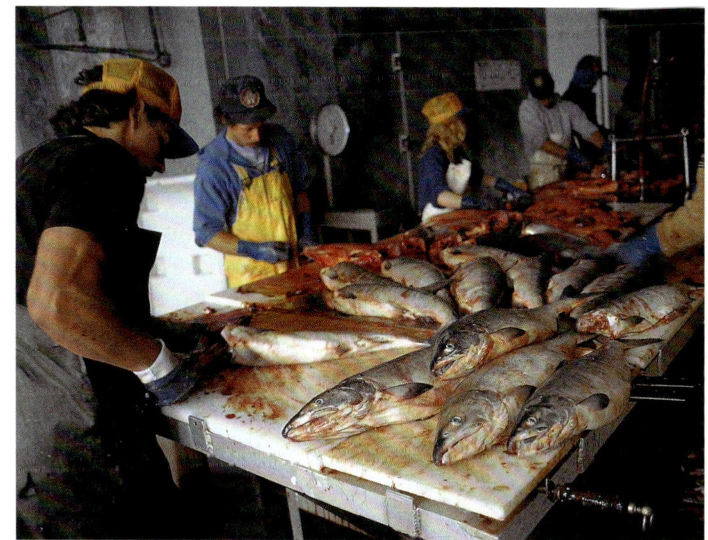

almost photographic representations of life in the sixteenth century. Ray especially liked Bruegel because he was elementally democratic in his selection of subjects and he was willing to do whatever was necessary to make a living as an artist. Bruegel painted for the masses, and his art was distributed in exquisitely detailed engravings. Shortly before him, Albrecht Dürer, another of Ray's art heroes, turned out beautiful woodcuts and engravings by the thousands. For the first time in the history of the planet, anybody could have a piece of work that had been intentionally created to be nothing more—and nothing less—than art.

"When I got to Alaska," Ray said, reaching over to turn down the volume on Steve Earle's "Transcendental Blues," "I saw other people doing the whole art and commerce thing that I idolized. It was a fertile time for people making art, and there were also the necessary people who understood it and would buy it. Bill Spear was up in Juneau making beautiful cloisonné pins and building a very successful business around them; Tom Sadowski and Jimmie Froelich were in Anchorage staging and photographing hilarious, interesting performance tableaus and putting them on postcards which sold like hot cakes; and a lot of other people were trying to do something other than paint moose, bears, and ice scapes. In Juneau, a group including Ken DeRoux, Dan DeRoux, Bill C. Ray, Jane Terzis, Mark Daughhetee, Paul Gardinier, and others started called themselves Arts R Us. In Anchorage, Mr. Whitekeys was bringing performance and visual art together in a unique commercial collision." Even in Ketchikan, there was—and still is—a thriving art scene that has supported and inspired Ray's work over the years. The ring of his occasional collaborators and playpals there includes Evon Zerbetz, Dave Rubin, Carla Potter, Lezli Morgan, Mary Henrikson, Terry Pyles, Halli Kenoyer, Hall Anderson, Chip Porter, and musician Russ Wodehouse. The native carvers and artists there were an *enormous* influence on him, including Nathan Jackson and his son Steve, Marvin Oliver, Israel Shotridge,

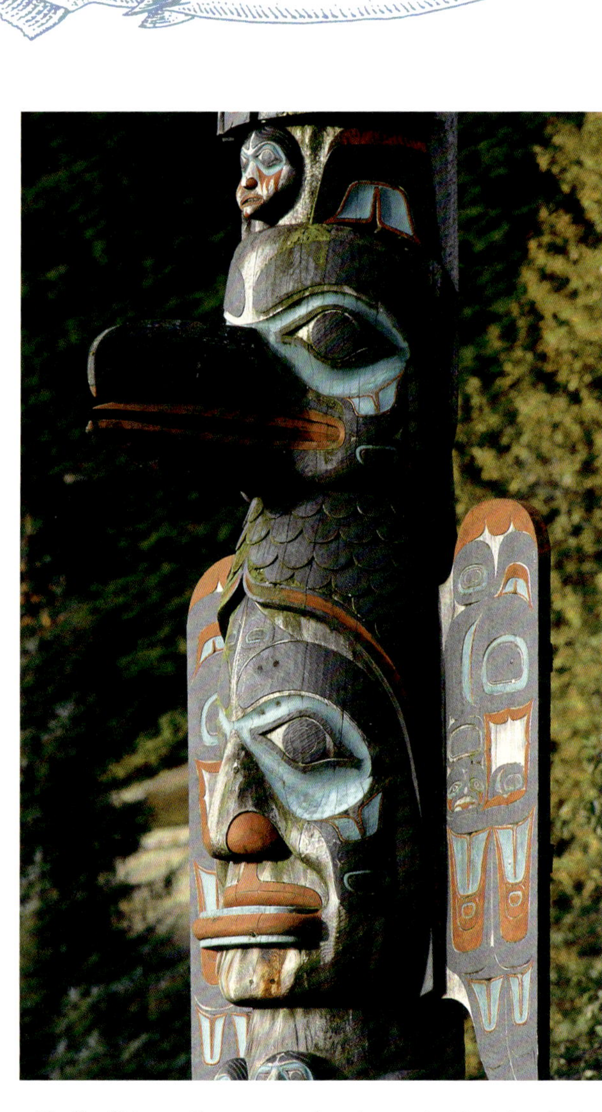

The Chief Johnson Tlingit totem pole at the mouth of Ketchikan Creek tells the story of the creation of salmon. This is the second version of the pole, carved by Ray's friend Israel Shotridge.

Donny Varnell, Lee Wallace, and Ernie Smeltzer. And people like Bob Waldrop and Terry Gardner were fish buyers in Ketchikan, but they also bought art. "It just felt like a good time to be there," said Ray.

The breeze blowing through the windows of Ray's studio also carried the ripe smell of fish from the packing plant chugging and steaming below. It was the most natural thing in the world for him to walk down to the dock and sketch a salmon. There was a marvelous array of profound imagery all over Southeast Alaska—totem poles, ravens, eagles, whales—and it would find its way into his work, but on that cannery dock in Ketchikan, Ray Troll took the turn that would define his life forever. Fish were present in his work as early as 1980 in "Fish Spirit Juju," and 1981 in "Plenty of Fish in the Sea," and even before as he worked through the totemic animal influences (from Gaylen Hansen on) as a Robot Slave Boy in Pullman. Those early Troll fish were purely iconographic, however, with little attention to scientific detail or accuracy, and a bit too earnest and transparent in their intention. All that changed, though, when he showed up in Ketchikan and began consorting with fishermen, fish mongers, and the fish themselves.

The T-Shirt King

As we blasted through the weird, moonlit panorama under the sleeping volcano, still two hours from Redding, Ray and I were getting a little bleary. I'd been driving all the way from Seattle, partly because I like to drive (Ray says I wanted to be the dad on the trip) and partly because he didn't want to surrender control of the CDs. At the moment the CD player was off, but we discovered that certain groove patterns in the concrete of I-5 create different ranges of tire hum, in which we had already heard the first few bars of Taj Majal's "Fishing Song," the Talking Heads' "Take Me to the River," and "America the Beautiful." That got us onto the idea of regrooving all the Interstates to play music in perfect tempo and pitch when you're driving the speed limit, eliminating the need for a lot of signs and cops, plus being cool beyond belief. Soon, though, we were back to fish.

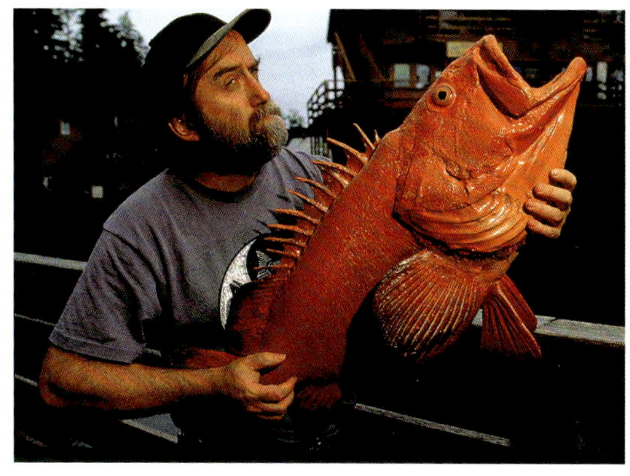

The artist and a plastic red snapper shamelessly mug it up for the camera on Creek Street.

"I really do love my finny brethren," Ray said over the tire hum, which right then sounded only like tire hum. "I know this artist who does mainly fishing boats at sunset. He and I were having a rambling conversation about stuff and he said, yeah, he'd taken the fishing boat thing and he was cashing in on it. He was surprised to find out that I was running with the fish thing but it wasn't merely me cashing in on it. I truly *love* the fish. The moolah thing was not the motivator at all. When I come across a new fish or critter, I just want to know more and more about it."

In Ketchikan, Ray's salmon, halibut, and rockfish soon became beautifully accurate renditions of the animals, but his imagination took them to some very unlikely and inspired places. By 1984, fish filled the air over people asleep in their beds beneath the sky of the Tongass Narrows, in front of Ketchikan. These fish hovered in Rousseau's jungle, cuddled up to a beautiful blonde (reminiscent of Michelle, who would become Ray's wife), and perched mouth down on the head of the artist. The work was whimsical, but rendered with painstaking accuracy, and people quickly caught on. The tension and offbeat cadences that Ray set up between real fish and human situations evoked laughter and wry enlightenment.

Ray was in the flow, completely swept away by his ichthyomuse, and everything was falling into place. He was out fishing one day, scuffing the bottom for halibut, when he hooked and landed a ratfish. These creatures are part of the shark line, cartilaginous, very strange looking, and ancient, and they began pouring from Ray's pens and brushes with the detail and accuracy that was becoming the hallmark of his work. The ratfish became his talisman and one of the sources of the wonderfully original humor in many of his drawings and paintings. Quite possibly, Ray drew the first pictures of ratfish that were not exclusively scientific illustrations. In one memorable drawing, "Cult of the Ratfish" (1985), Ray depicts three men wearing ratfish on their heads approaching a nude woman also wearing a ratfish headdress. Ratfish haven't left Ray's consciousness for long ever since. They are the seeds of his later fascination with sharks, his *nom de soirée* is Ratfish Ray, and his Alaska vanity license plate is RATFSH. He formed a band called the Rapping Ratfish Brothers and played at fisheries society meetings. A few weeks before we were humming along the highway in the Taurus, one of the world's leading experts on ratfish, Dr. Dominique Didier Dagit, described a brand-new species and named it after Ray, *Hydrologus trolli*. Dominique said in her announcement of the new species, "It's kind of nice to name a species for someone, and I thought, 'Here's a chance to name one for someone who's really interested.' It kind of looks like him, but with less facial hair."

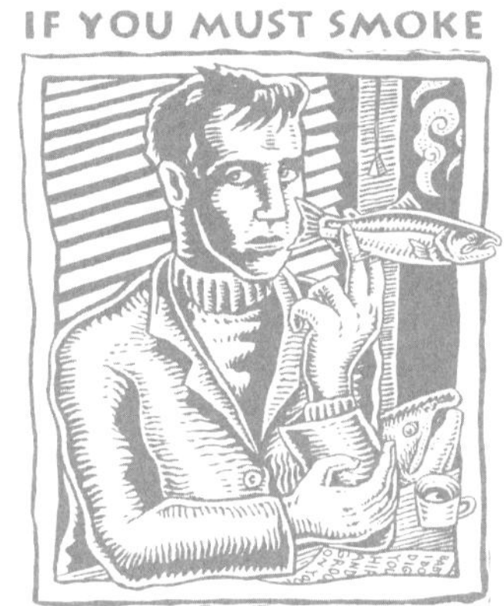

As Ray's fish madness was gaining momentum in 1984, something happened that admitted him to the ranks of the Bruegels and other artists who manage to work with originality and integrity and still make a living. On a whim he screened the words "Let's Spawn" on three hundred T-shirts for a Ketchikan seafood festival, then proceeded to sell every one of them in two days. After that, Ray realized that the market of summer fishy folk in Ketchikan alone made the shirts worth his while, and he quickly issued "Humpies from Hell," which immediately became a mandatory accessory for anyone who had ever spent time on a slime line cleaning pink salmon, a.k.a. humpies. "Spawn Till You Die," in which a pair of salmon form the crossbones in a classic death's head, became a huge seller and the image for which Ray believes he will be most remembered when he's dead. After just a couple of years in his garage-based T-shirt venture, with an expanding clientele of fishermen and tourists, who spread his shirts around the country like spores, Ray set up the Troll Line as a real business in partnership with a screen printer in Seattle. He quickly joined contemporary Gary Larson in the forefront of the bizarre, irreverent, enigmatic, and distinctly Pacific Northwest brand of humor rooted in animals and natural history. The slogans that made the shirts so funny, wry, or insightful now became critical elements of his creative process. Some, he says, are just stupid puns and jokes that he couldn't resist; others are more philosophical. Many have worked their way into the vernacular: "Spawn Till You Die," "Bassackwards," "Ain't No Nookie Like Chinookie," and "Fish Worship: Is It Wrong?"—to name but a few. Under the pseudonym of Rollo Trout, Ray even descended into the utterly lurid when such jewels as "The Perils of Nude Fly Fishing," "Northern Exposer," and "Waiting with Baited Breasts" made their way into the collection. At one point a publisher was considering a collection of Rollo Trout images, but then Ray's father happened to see one of them on a T-shirt in Juneau, at which point he delivered an honest-to-goodness lecture to his successful son about what such depravity could do to his career. A museum director confirmed Colonel Troll's assessment, so Rollo Trout and his oeuvre now languish in relative obscurity.

In the eighteen years since "Let's Spawn" brought down the house in Ketchikan, Ray has sold something like two million T-shirts. His images have also appeared on coffee cups, coffee bags, mouse pads, screen savers, books, lunch boxes, pencil cases, card games, baseball hats, magnets, an Easter egg commissioned by the Ronald Reagan White House, greeting cards, magazine ads, billboards, and postcards. (Ray thinks postcards are a great way to throw art at the world.) Ray has drawn on skin, and some people have his images tattooed on themselves. Harrison Ford was spotted wearing a "Spawn Till You Die" baseball hat; in a scene in the movie *Safe*, Julianne Moore gazes at Ray's Greenpeace poster; a photo in *National Geographic* featured a scientist wearing his "Out of the Ooze and Born to Cruise" T-shirt; and members of Mötley Crüe and Led Zeppelin have worn "Spawn Till You Die" T-shirts on stage.

"Commerce can be a devil," Ray said after we'd run through the catalog of places his art has appeared. "There have been a couple of moments when I was totally fed up with it, but the fun of interacting with an audience always brings me back. And I like to think I'm continuing the tradition started by Bruegel; everybody should own art, even if it's on underwear."

And then the day was over, with 608 miles on the trip meter. We checked into the Best Western and ate iceberg-lettuce salads with French dressing, prime rib, and mashed potatoes at Lulu's. Ray called Michelle and they discussed Corinna's new blue hair; he was talking to Patrick about his basketball game that night when I left to ice the cooler. I came back, gave him the TV remote, and he flipped through fifty-seven channels, finding nothing on but a rerun of a Rolling Stones concert when the lads were in their fifties. I dropped off to sleep with the mental image of a sneering Mick singing "Sympathy for the Devil." I was glad to be the first asleep because after thirteen years of road trips with Ray, I knew that his snoring is like the soundtrack from *Battle of the Water Buffaloes.*

Stream of Consciousness

Ray and I met on a blind date in 1990, arranged by Marlene Blessing, an editor who thought we should collaborate on the book that became *Shocking Fish Tales.* I had heard about him: Ray had been writing about fish, fishing, and Alaska for a few years by then, and he'd amassed a substantial portfolio of images, becoming something of a celebrity in Alaska, the Pacific Northwest, and everywhere fishermen gathered to lie about their luck. His work featured not only the superb draftsmanship of a science illustrator but also the wit and whimsy of an immediately accessible humorist. Moreover, his major works, including "Rapture of the Deep," "Dance of the Fish Charmers," "Midnight Ritual," "Daughter of Fog," "One Shot Jack," "Assault from the Air," and "Rain on the Parade," took Ray's totemic iconography—and his fish—into some adventurous and beautiful territory. As one California admirer put it, "This isn't about fish, it's about the human condition."

Our first meeting took place in my apartment overlooking Lake Union in Seattle. Ray arrived with a tray of slides; we circled each other for a while, finally sat down to look at pictures and talk, and didn't get up again for six hours. The next day both of us called a mutual fisherman friend in Ketchikan and asked, "Hey, Barre, who is this guy?" Neither Ray nor I came without some liabilities and questionable decisions in our pasts, but after we'd run each other's references, we decided to give collaboration a try. The arrangement between a writer and an artist is usually that of reporter or storyteller and illustrator, but in our first meeting we decided to dump that paradigm in favor of a different kind of relationship.

The idea was that we would begin with our shared fascination with fish, then make a selection of Ray's images, which I would complement with a series of short essays, and together come up with new essays and images that rose from our collaboration. The book would be fueled by the fish themselves, our senses of humor, and the shared belief that science and fun are not mutually exclusive. Our job, we agreed, was to become fascinated with our subjects and, with whatever skill we could muster, share that fascination with other people. Both of us believed that folks who laugh, are curious, and have a good time are much less likely to be mean to each other, start wars, foul their nests, and generally make life miserable. From the beginning, we envisioned a book that was a kind of Trojan horse filled with excitement about life—particularly of the sea and its creatures. Our greatest satisfaction, the mark of the book's true

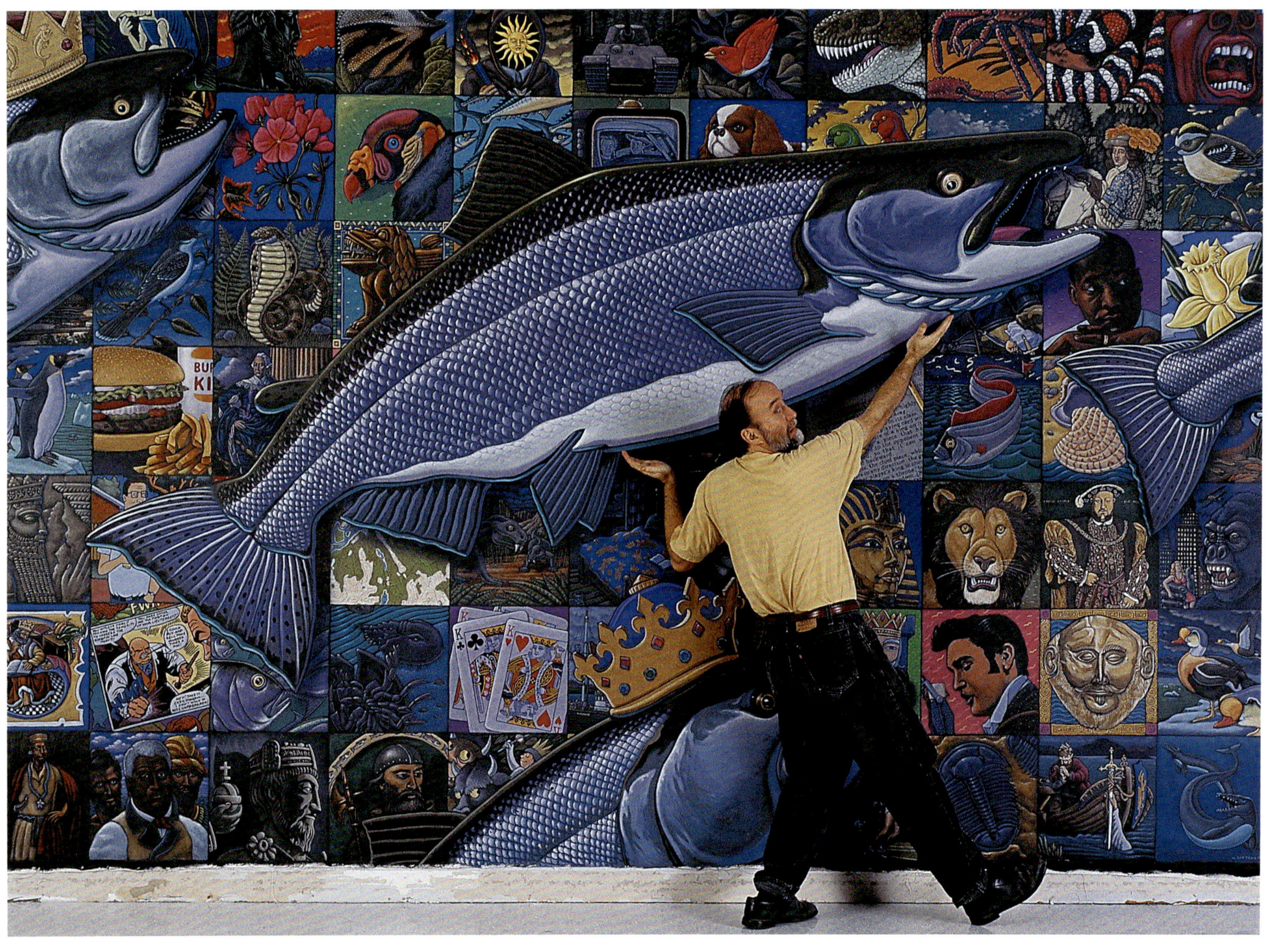

Ray spent most of 1998 painting this large mural, Kings, *for the local high school in Ketchikan. The K-High mascot is the king salmon, so he riffed on all the names, places, and things that are called kings. The painting, made up of sixty-eight squares, has everything in it from Elvis to trilobites named* Elrathia kingii.

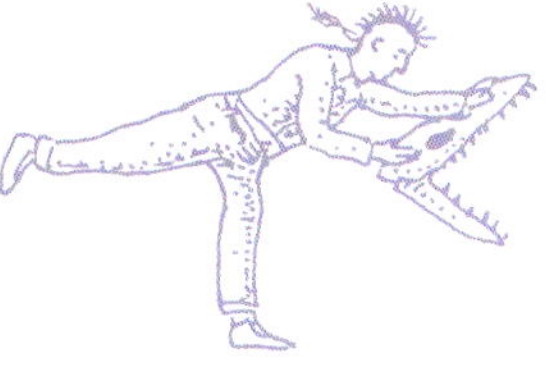

success, we figured, would come from seeing the book on the dashboard of a pickup truck parked dockside.

We also planned to have a good time, and the first order of business was obviously a road trip. So off we went to a little community marine center in Port Townsend, Washington, to look at fish and spend some time getting to know each other. When we walked through the door, the curator took one look at us and said, "Brothers, right?" He then showed us around. We drove back to Seattle, and a few days later I wrote my first essay for the book, inspired by a tank of Pacific spiny lumpsuckers. Ray drew a picture of these strange little fish, and we were off and running.

We talked on the phone almost every day for six months, faxed words and pictures back and forth, wrote and drew and laughed, and formed opinions about each other. I'll have to wait until he writes my biography or delivers my eulogy to know what he thinks of me, but I can tell you here and now that Ray Troll is a *mensch*. That's the most accurate word I know to describe him. It's Yiddish for a person having admirable characteristics such as fortitude, firmness of purpose, and a fundamental decency—a real human being. The first thing I liked about Ray was that he knows how to work, knows how to show up and get things done every day. I liked how he was around his kids, genuine, kind, and funny, but always Dad. I liked how he and Michelle took care of business but always seemed to be sharing a good secret that made them happy to be together. I liked how Ray treated his army of friends, how he obsessed on seeing as many of them as possible when he passed through their towns. I liked how he encouraged other artists with his friendship and connections and helped them thrive, regardless of the level of their talent or achievement. I liked it that he had moments when he lapsed into an earnestness that rendered him as vulnerable as a child.

Shocking Fish Tales came out in 1991. We had intended to call the book *Fish Worship: Is It Wrong?*, but at some point GTE, the Bible Belt–based corporate parent of Alaska Northwest Books, got wind of the somewhat heretical title and summarily killed it, giving Ray and me our first ride in the publishing business dunking chair. Delicious irony struck a few weeks later, though, when the PR person unveiled the ten thousand bookmarks she'd had printed for our promotional tour. The florid pink piece of cardstock had a salmon head on one end and looked for all the world like a penis, adding yet another aspect to the democratic celebration of the art of Ray Troll. To our pleasure, the book did show up on the dashboards of pickup trucks, and it's still in print.

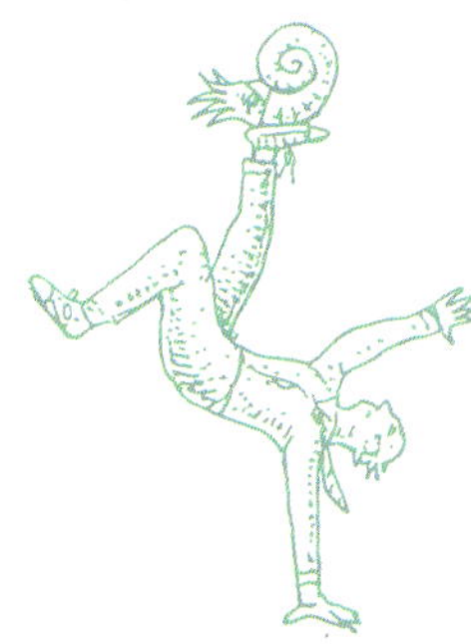

Dancing to the Fossil Record

The run from Redding to Monterey was a piece of cake, a cruise-control streak on straight, flat highway that took five hours, including a stop for breakfast at a diner decorated with chain-saw bear sculptures. Back on the road, Ray decided it was time to introduce me to the Flaming Lips ("Three guys from Oklahoma nobody ever heard of before . . .") as we rolled through the Great Central Valley, dazzled by the endless rows of fig and nut trees flashing in perfect alignment no matter which direction we looked.

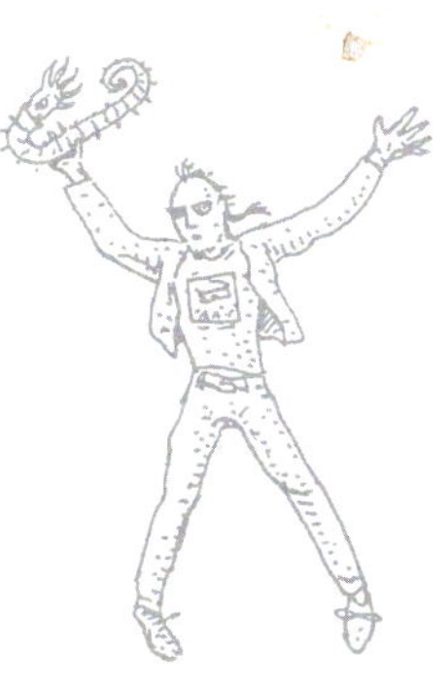

The Monterey Bay Aquarium is one of the world's great Fish Temples—the perfect place, it seemed to us, to celebrate twenty years of art by Ratfish Ray. We could visit some friends, hang around the aquarium (staring at fish in big tanks is soul mending for both of us), and see what developed, which has been pretty much our MO since we started working together.

As we were finishing up *Shocking Fish Tales*, for example, we stumbled across the fact that the ancestors of all land vertebrates—including humans—lived in the sea. This was wonderful news. Ray and I, it turns out, are not just fish worshipers; we *are* fish—with highly modified fins, but fish nonetheless.

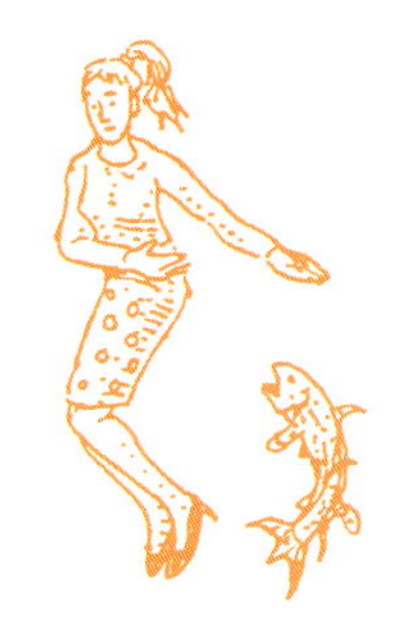

There was never any question that we would do another book together, and now the next project became clear: we would trace the evolution of fish. This time, though, it wouldn't just be a retrospective of Ray's work and a few light, funny essays of mine. First we would travel back in time through fossil beds and the

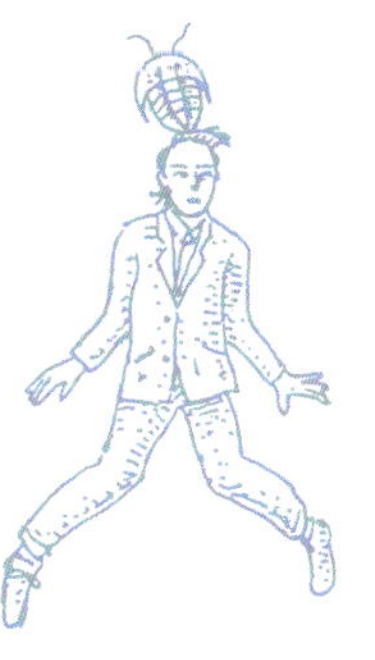

minds of paleontologists, ichthyologists, and other scientists to discover our most ancient roots; then we would start from scratch with words and pictures. We were more than a little bit frightened. It was collaboration without a net for both of us, and for Ray especially it represented a pretty significant risk. His fish print and T-shirt thing was clicking along like a well-oiled machine, serving him creatively as well as practically. Taking off into rocks, fossils, and the meaning of life might be the equivalent of jumping off a cliff into artistic oblivion, not to mention abject poverty.

We built the book around road trips, of course, one a month-long meander through the legendary fossil beds of Kansas; Green River, Wyoming; and the Red Deer River Valley in Alberta, tracking fish, dinosaurs, birds, and other creatures of the distant past now speaking to us about the origins of life. We went to museums and aquariums in Monterey, Los Angeles, Seattle, New York, and Washington, D.C., where we discovered the magic words for getting into the paradise of their warehouses and back rooms: "We're writing a book." Ray filled sketchbooks, I filled notebooks, and one day when we were walking through the Los Angeles County Museum of Natural History we hit on the title for our book: *Planet Ocean—A Story of Life, the Sea, and Dancing to the Fossil Record.* Marlene Blessing signed on to edit, our *Shocking Fish Tales* designer, Kate Thompson, offered her talents, and our lives again blossomed into delight.

Being around Ray while he sketched, drew, and painted was more transformational for my own work than the two years I spent getting an MFA in writing. I watched him organize the imagery in a way that was very new to me, beginning often in one corner of the frame and filling it with lines and colors that seemed simply to suggest themselves as he went along but eventually yielded a cohesive, elegant result. Until *Planet Ocean,* I didn't fully realize that I too could work with the assurance that

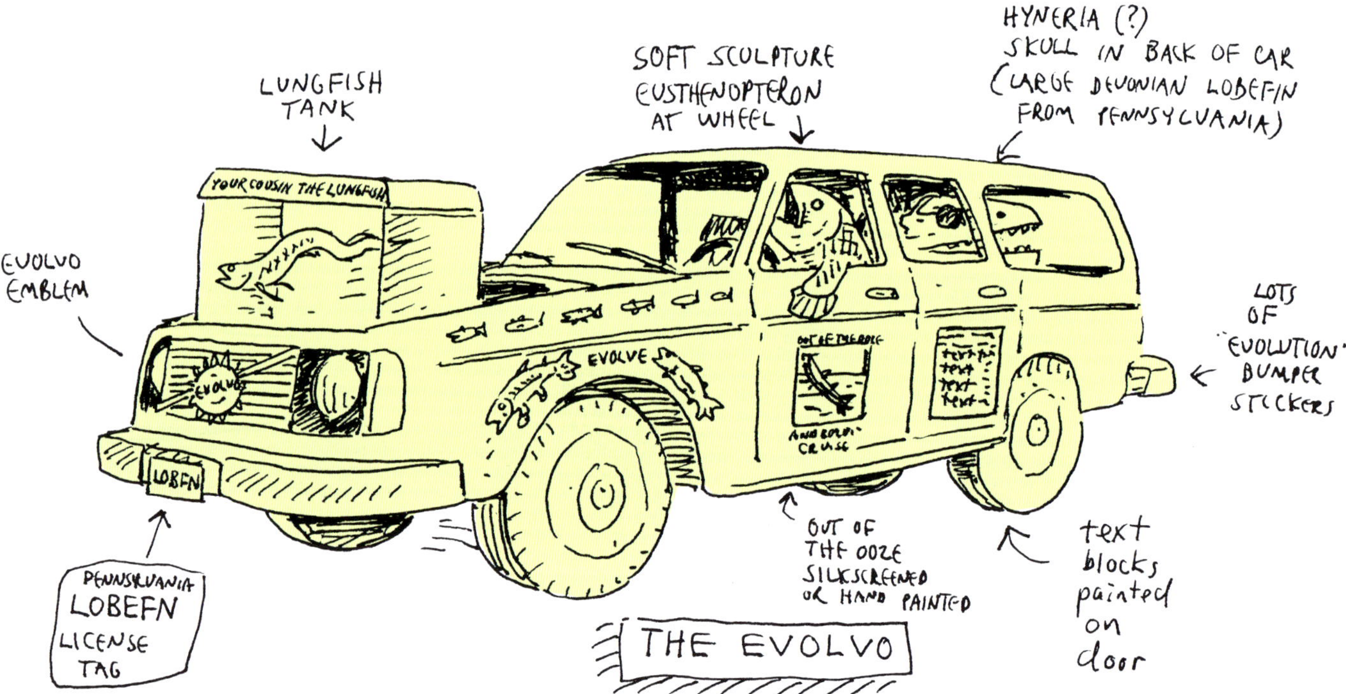

the process would lead me to an original, often unpredictable outcome. Before then, I assembled stories; in *Planet Ocean*, I allowed them to happen. Ray got into a new kind of groove on the road trips, too, steadily filling sketchbooks and creating fully realized paintings and drawings inspired by our journalistic inquiry. My orderly methods of gathering information brought focus to what he was encountering without inhibiting his visual impulse.

Planet Ocean was published by Berkeley's Ten Speed Press in 1994, and we followed it two years later with a kids' version, *Raptors, Fossils, Fins, and Fangs*, from Ten Speed's children's subsidiary, Tricycle Press. Meanwhile, the celebrity Ray already enjoyed among ichthyologists, fishermen, and slime line workers now spread through the world of geologists, paleontologists, and evolutionary biologists. We made a few mistakes in telling the story of life, the sea, and the fossil record, but none were bad enough to send scientists running away from us.

Soon after *Planet Ocean* came out, Dr. Peter Ward, a paleontologist then at the University of Washington's Burke Museum, suggested we turn the book into a natural history exhibit. Again, Ray Troll's art was hurtling toward the masses in yet another form. For both of us, combining art, words, humor, fossils, live animals, and whatever else we could get our hands on to tell a difficult science story was pure heaven. The first exhibit based on *Planet Ocean* opened at the Burke Museum in the summer of 1994. The show was then variously reincarnated at major museums in Juneau, San Francisco, Philadelphia, and Denver and at the Oregon Coast Aquarium in Newport. Ray did all the hard work, filling the walls with ghostly shadows of animals past and present, hanging his original works from the book, gathering fossils from individuals and museums around the country, and directing the endless details of such an installation. As the artist William Wiley punned on opening night at the California Academy of Sciences in San Francisco, Ray left "no tone unturned." There were giant sculptures of saber-toothed salmon and sharks, and interactive toys for children. Shameless visual puns included a fossil bed (involving a bunk bed) and an "Evolvo," an

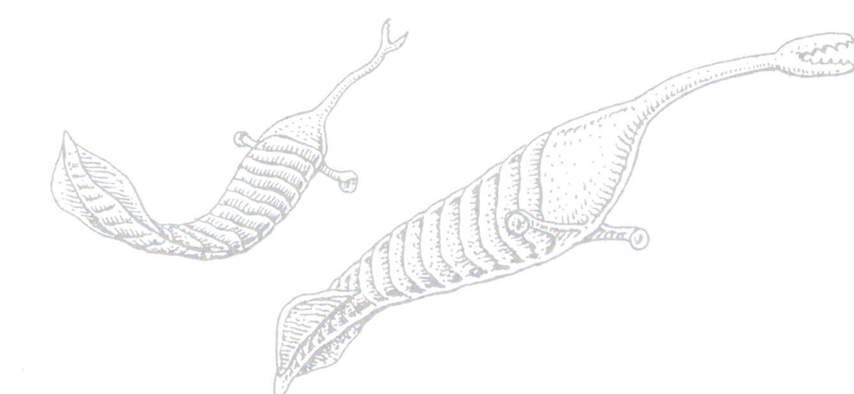

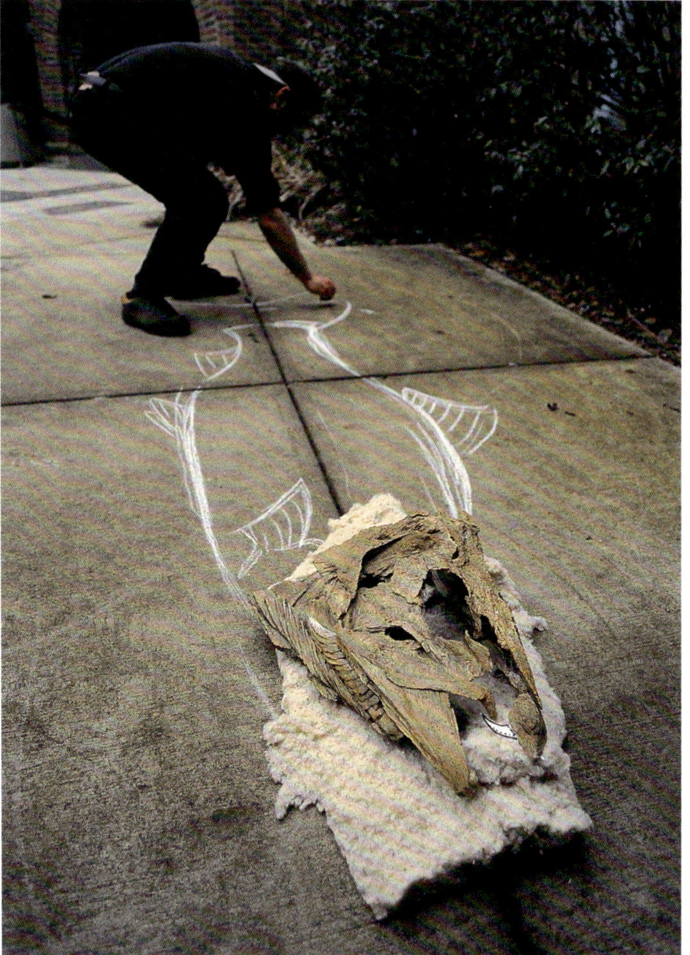

Ray sketching the body outline of a saber-toothed salmon at the University of Oregon in the spring of 1997. "I made a pilgrimage to see the holotype (the first specimen ever found) in their collection, and to my surprise they let me cart it out onto the sidewalk where I could get a better sense of how big these extinct fish were. This one must have been about six feet long. I've seen bigger specimens in other collections, and scientists estimate they grew to about nine feet, a whopper in any salmon fisherman's wildest dreams."

actual Volvo station wagon decorated with fossils and slogans, complete with Charles Darwin at the wheel. Ray commissioned a music soundtrack that debuted in San Francisco, featuring his pal Russell Wodehouse from Ketchikan, who wrote the songs and played most of the instruments, and he recorded assorted scientists rapping on the story of life in the sea. By the time the show closed in Denver in 1999 it had played to half a million people.

Fish Storm on the Amazon

After the *Planet Ocean* exhibit placed Ray firmly in the lineage of great explainers of real science, he embarked on a collaborative adventure of a different kind. It all started in 1997, when he and I spent three weeks on the Amazon; Ray returned there in 2000. So far several drawings and one major mural have resulted from his experiences in the Amazon. More significant, though, is the way the artistic/scientific perspective has been enlarged in the field: each of Ray's two trips on the big river included geologists, ichthyologists, botanists, photographers, visual artists, filmmakers, writers, and even environmental policy types, all thrown together on one boat for weeks at a time. The trips created a network of continuing interaction, chiefly via email, and the outcome will likely be a major touring exhibition and at least one book, which Ray hopes will increase awareness about the Amazon basin and its creatures, an awareness that is desperately needed as that region is sapped of its health and integrity by an increasingly abusive human presence.

For Ray, of course, the Amazon is fish paradise. One night, we traveled by outboard canoe close to the riverbank after dark, shining flashlights into the shallows. A storm of frightened fish rose up, including a two-foot arowana that gave one woman a black eye, and a bright peacock bass that buried itself in another's thigh. The storm was so intense at times that we had to use our seat cushions as shields against a nightmarish swarm of toothy, spiny, wriggling fish. Back at the boat we collected seventy or eighty carcasses from the canoes, enough for days of meals. Most evenings on the river were spent telling piranha-bite stories and listening to impromptu lectures about the sky, the plants, the fish, and our lives in the context of it all. After the Amazon, Ray redid his business cards to read, "Ray Troll: Art, Science, and Mystery."

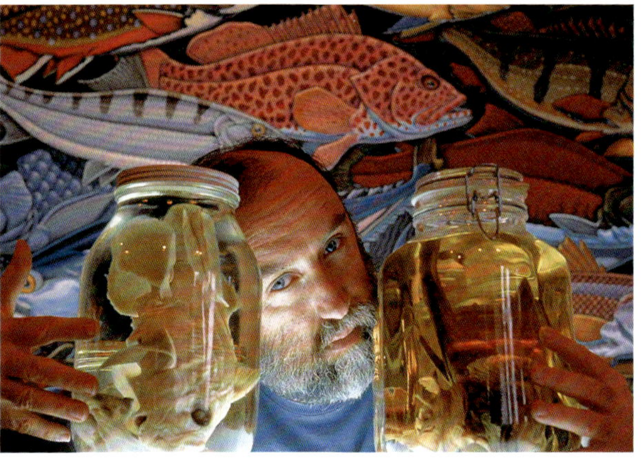

The artist in his studio with a few choice pickled specimens.

A Sea of Sharks from A to Z

And then there are the sharks. *Planet Ocean* brought another unexpected outcome, steering Ray's work into a new realm of curiosity, joy, and beauty. One afternoon in the basement of the L.A. natural history museum we were with our pal, paleoichthyologist Dr. J. D. Stewart, when he bent down and pulled a stumplike hunk of gray-black rock out from under a storage shelf. He dusted it off and moved it into the light. On the top surface of the rock we could see an impression that looked a lot like a big snail shell or chambered nautilus. "Check this out," J. D. said, "It's blown paleontologists' minds for years."

The mysterious spiral was the tooth whorl of an extinct sharklike fish called *Helicoprion* that lived in the Permian ocean about 285 million years ago. To Ray, it was like bait dangled in front of hungry trout: he went for it and got hooked good. That fossil led Ray to Dr. Rainer Zangerl, an authority on Paleozoic sharks who, easing into his eighties in Indiana, burned with passion and curiosity for ancient ocean creatures. Ray went to visit Dr. Z, wrote his own section of *Planet Ocean* on his newfound favorite critter, and never let go of sharks.

Nobody had figured out what *Helicoprion* or a number of other extinct sharks looked like because, being cartilaginous fish, sharks usually leave only teeth behind when they die. Ray got busy building a network of shark experts, though, and after many false starts came up with a pretty good idea of what the whorl-toothed monster might have looked like. And so he became the first person to make a believable drawing of *Helicoprion* and its weird dental apparatus. This eventually led to an appearance on a Discovery Channel special on ancient sharks with American shark guru Dr. Richard Lund. Ray has since been a featured speaker at conventions of the world's leading shark researchers, published a children's book called *Sharkabet: A Sea of Sharks from A to Z,* and staged a traveling shark exhibit that at this writing has already been to Minneapolis; Anchorage; Seattle; Bozeman, Montana; Ketchikan; and Mesa, Arizona.

Doors have opened to Ray throughout the shark tribe, including that of Dr. Gregor Cailliet, who lives on a hill overlooking Monterey Bay, where we showed up on schedule early one afternoon in January 2003. Greg had offered us his hospitality for a couple of days, so we unloaded the car and said hello to a room full of people settling in to watch the NFL playoff game between the Oakland Raiders and another team that was lost to us because neither Ray nor I is a football fan. Still, it looked like a good time was shaping up because one of the Raider rooters ran a shark research foundation, another was a graduate student of Greg's at nearby Moss Landing Marine Labs, and another was Bob Lea, an ichthyologist practicing on the California coast for almost forty years and coauthor of the definitive guide to local fish. Ray and I were a little road weary but open for business, so we added what was left of our barbecue to the assortment of shrimp, guacamole, and chips on Greg's table, got a couple of beers, and sat down in front of the television set.

Midway through the second half of an Oakland blowout, we learned that a fisheries observer had just delivered a rare (for central California) sleeper shark to Bob Lea's lab. Bob asked if we'd like to come by the next day and take a look; the answer, of course, was yes. After we made the shark date, artist Roberto Salas showed up. He and Ray had collaborated on murals at the Monterey Bay Aquarium and at several of the natural history museums during the *Planet Ocean* run; they were now planning a new one for the National Marine Fisheries lab in Santa Cruz, so they had some business to do. This was a typical off-the-road day for Ray: a rush of plans, getting together with friends, meetings with sources for this or that science topic, a few beers, some dinner—and then it would start all over again the next day.

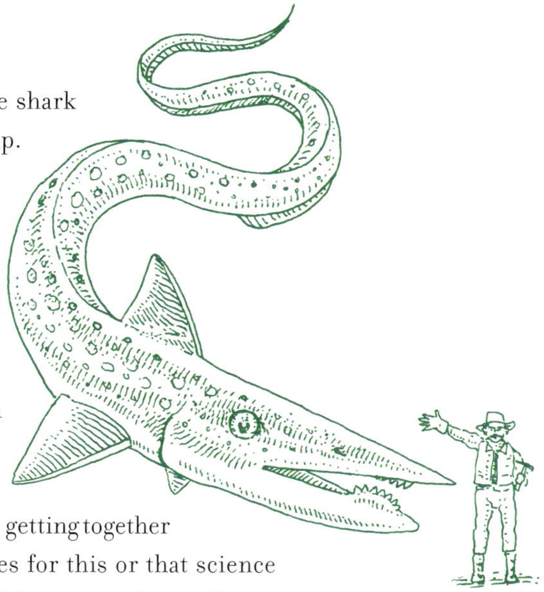

The next morning at Bob Lea's lab, Bob led us into a cluttered, utilitarian room with a drain in the floor, and he and Ray suited up in gloves and aprons to examine the shark. Ray had seen lots of sharks, dead and alive, but never a sleeper shark, so he was vibrating with excitement as Bob laid the brownish carcass on a stainless steel table.

Ray's rock-and-roll raps pale by comparison to his fish raps "*Somniosis pacificus,*" he said, naming the six-foot beast as he ran his hand down its flank. "They are basically dogfish, but they get bigger than great whites and eat whole fish, big chunks of seals, sea lions, and whales. Somebody once found a whole reindeer in a sleeper shark." Ray looked over at Bob, who by then had his entire forearm inside the shark's mouth trying to extract a small orange fish that was stuck in the shark's gills. Bob nodded at Ray with an amazed smile on his face. "Look," Ray said, touching a line of black pinpoints on the shark's skin. "These are the ampullae of Lorenzini. They're like direct lines that send sensory information to the shark's brain. Amazing. And man, look at those teeth." Bob pulled his slightly bloodied arm out of the shark and the two of them examined the mouth, which had teeth for grasping on top and teeth for cutting on the bottom. "Bob," Ray chortled, "you got bit by a dead shark!" Bob checked the wound; it was just a little nick, but true, he'd been attacked by a dead shark. "God, I love these guys," Ray sighed. He and Bob hauled the shark outside onto the loading dock where the light was better for a few pictures, and then it was time to go.

Bob told Ray what his favorite T-shirt was, gave us his card, and said, "Call anytime." In the car, Ray mentioned that it sure was a lot easier and more fun getting into labs and museums these days than it was fifteen years ago, which is about as close to pride as he gets when it comes to his accomplishments in science. He has a ratfish named after him, but I think one of these days he should get an honorary Ph.D. acknowledging his role as one of the great science teachers of all time.

No End of the Road

This is a book about Ray Troll's art, not a blow-by-blow account of two middle-aged guys driving two thousand miles in a Ford, and I want to let you move on to the pictures. You've got a pretty good idea of who Ray Troll is by now, so I'll just say that we left Monterey with no permanent injuries from a reception for Roberto Salas's art show at the Moss Landing Marine Labs, after which we retired to the Elkhorn Yacht Club, where the party really got going. The next morning we spent a couple of hours at the aquarium gift shop signing books; I was impressed, as always, by the attention Ray pays to people on the retail end of his art. We spent that afternoon aboard the Monterey Bay Aquarium Research Institute's research ship *Western Flyer* with the chief pilot of the ROV (remotely operated vehicle) *Tiburon*, Buck Reynolds, who had been a huge fan of Ray's art since he got a T-shirt from a daughter living in Alaska. We were awed by the thought of piloting the *Tiburon* down to ten thousand feet in the Monterey Submarine Canyon and sneaking up on a giant squid or an anglerfish. We finished the day by driving to Berkeley to see some more friends and do a little business, then drove back to Seattle, listening to the concrete of I-5 and imagining that it was singing to us.

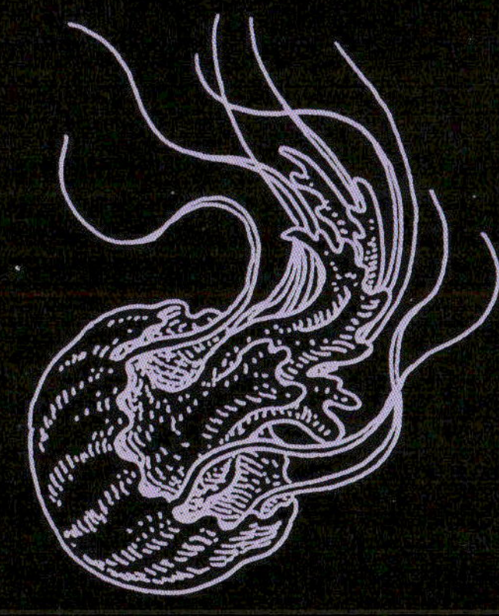

THE ART OF RAY TROLL

WITH COMMENTARY BY THE ARTIST

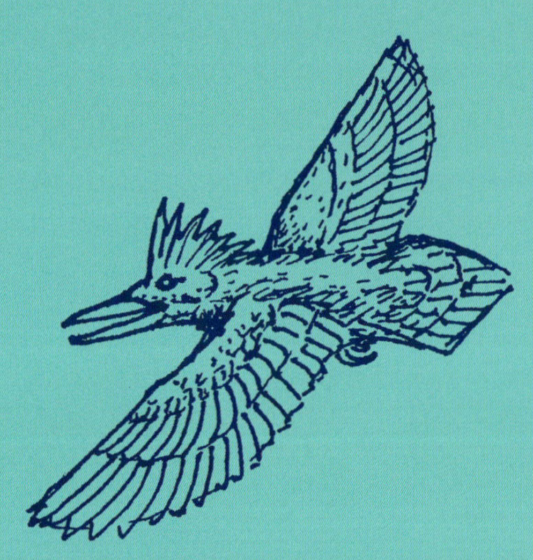

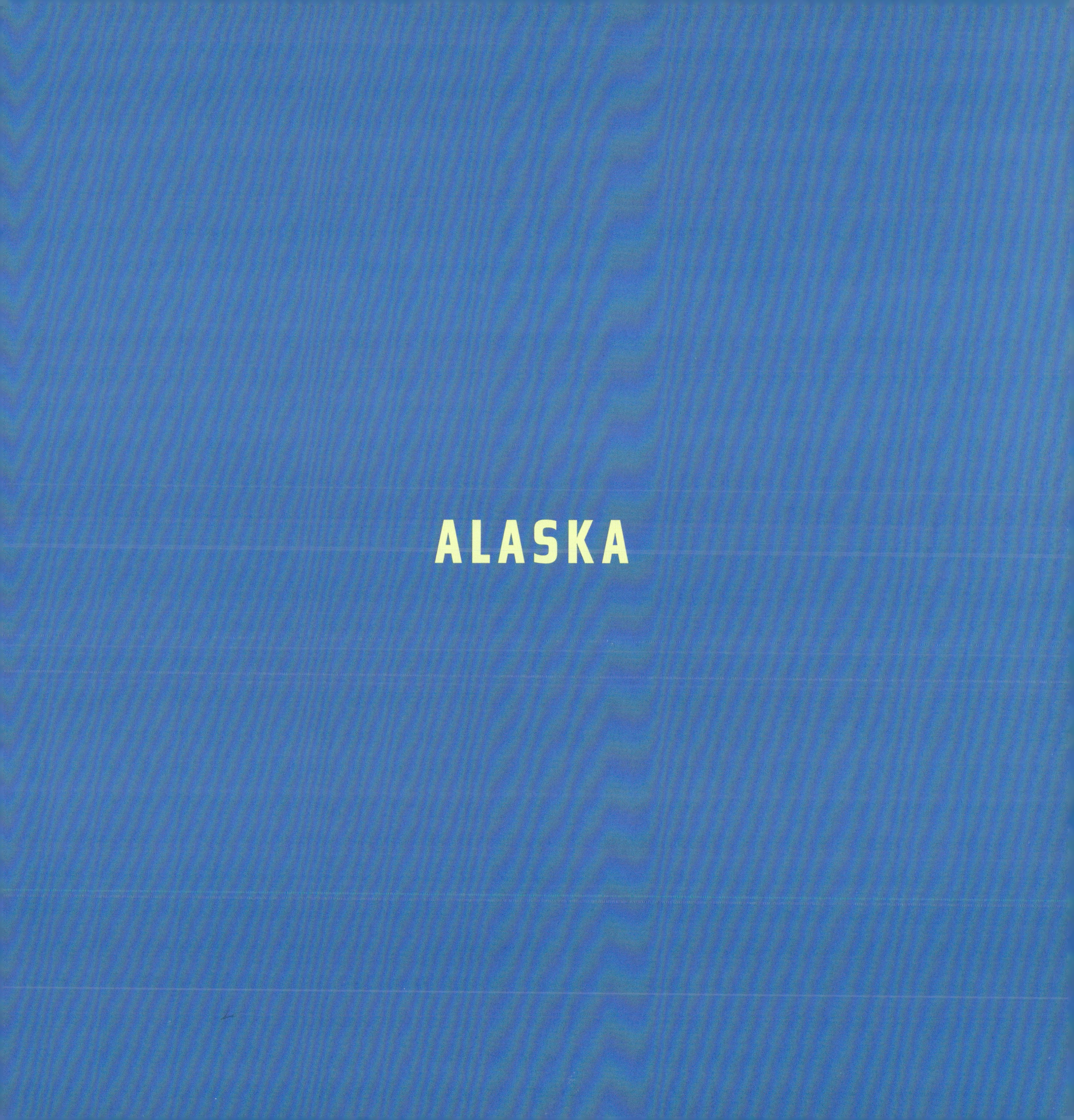

ALASKA

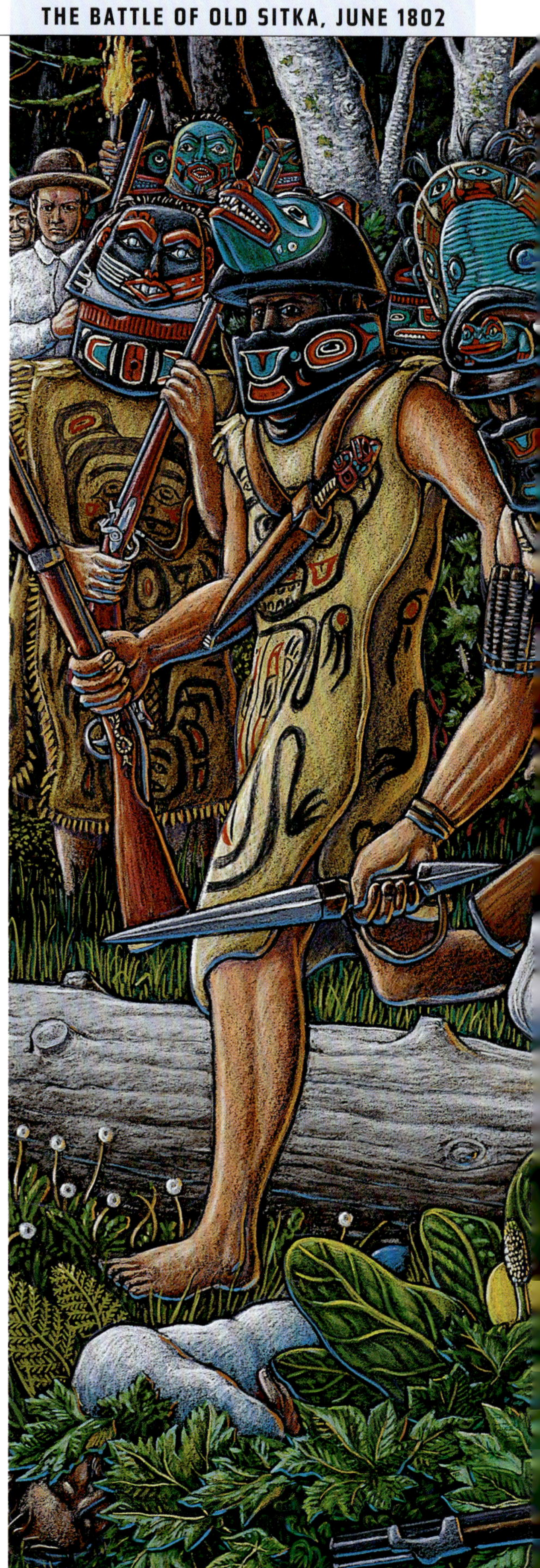

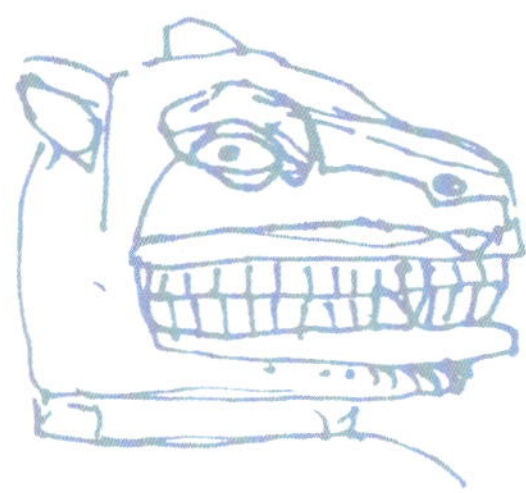

Sometime near summer solstice in 1802 hundreds of Tlingit warriors attacked the Russian/Aleut settlement of St. Michael's, near present-day Sitka, Alaska, killing nearly all of the inhabitants. In the carefully planned assault, half of the attacking forces came by canoe and the other half descended on the fort from the surrounding forest.

A Russian historian named Kiril Khlebnikov wrote this passage about the attack: The Tlingits "suddenly emerged noiselessly from the shelter of the impenetrable forests, armed with guns, spears, and daggers. Their faces were covered with masks representing the heads of animals, and smeared with red and other paint; their hair was tied up and powdered with eagle down. Some of the masks were shaped in imitation of ferocious animals with gleaming teeth and of monstrous beings. They were not observed until they were close to the barracks; and the people lounging about the door had barely time to rally and run into the building when the [Tlingits], surrounding them in a moment with wild and savage yells, opened a heavy fire from their guns at the windows. A terrific uproar was continued in imitation of the cries of the animals represented by their masks, with the object of inspiring greater terror" (quoted in Polly Miller, *Lost Heritage of Alaska*, 140).

The masks Khlebnikov refers to actually were heavy wooden war helmets worn on top of the head in combination with neck protectors; wooden slat armor or thick elkhide covered the rest of the warrior's body. When I first read this account of the battle I knew that I had to someday attempt a drawing of it.

In June 2002 I made my way to Sitka for the two hundredth anniversary of the battle. I went to the empty beach and tried imagining what the scene must have looked like. I then spent countless hours researching the helmets and armor before I started the drawing that July. To my mind, the Tlingit warriors bear a striking resemblance to Japanese samurai warriors.

2002, colored pencil on paper, 22 in. × 30 in.

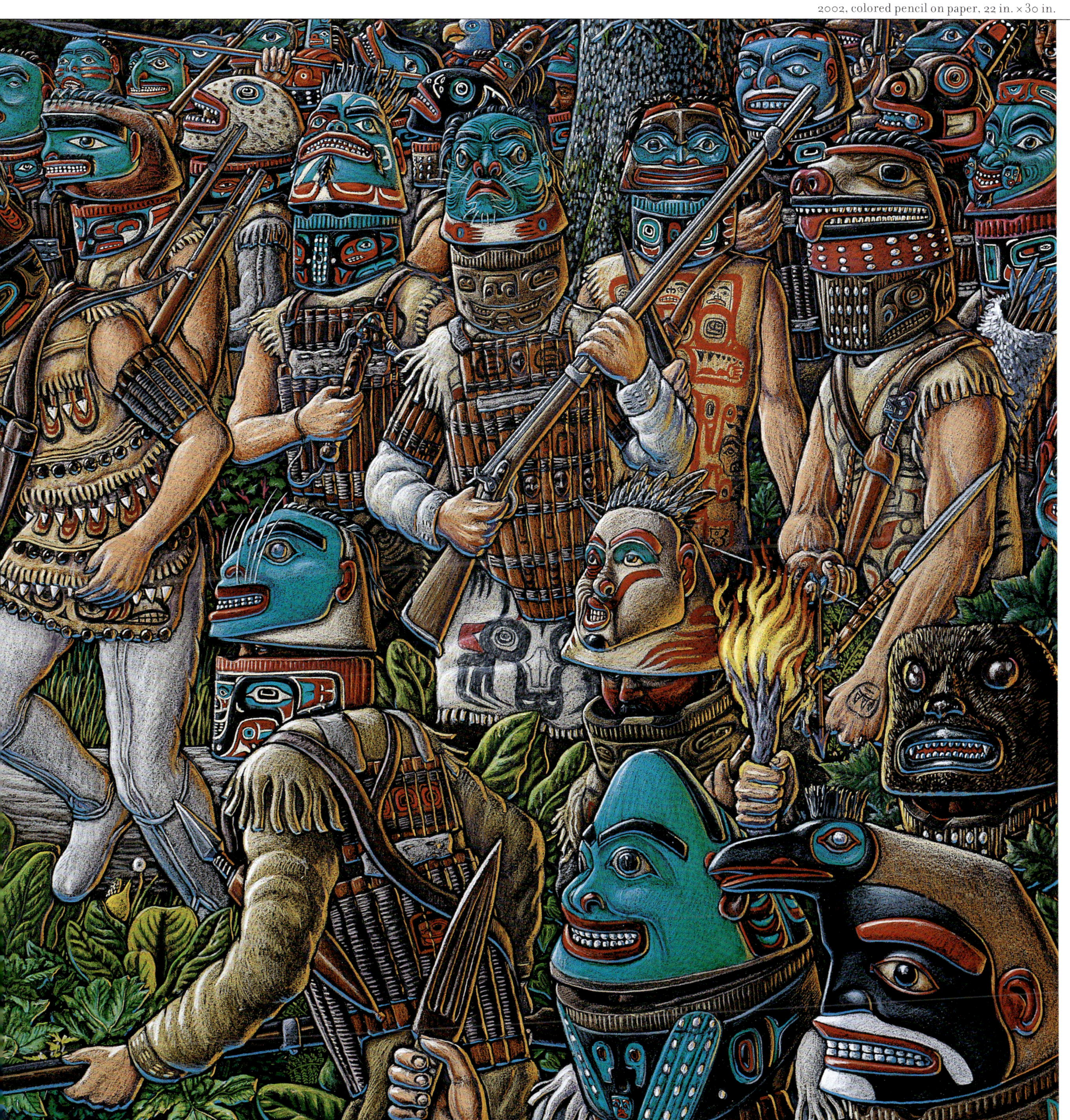

1987, colored pencil on paper, 14 in. × 19 in.

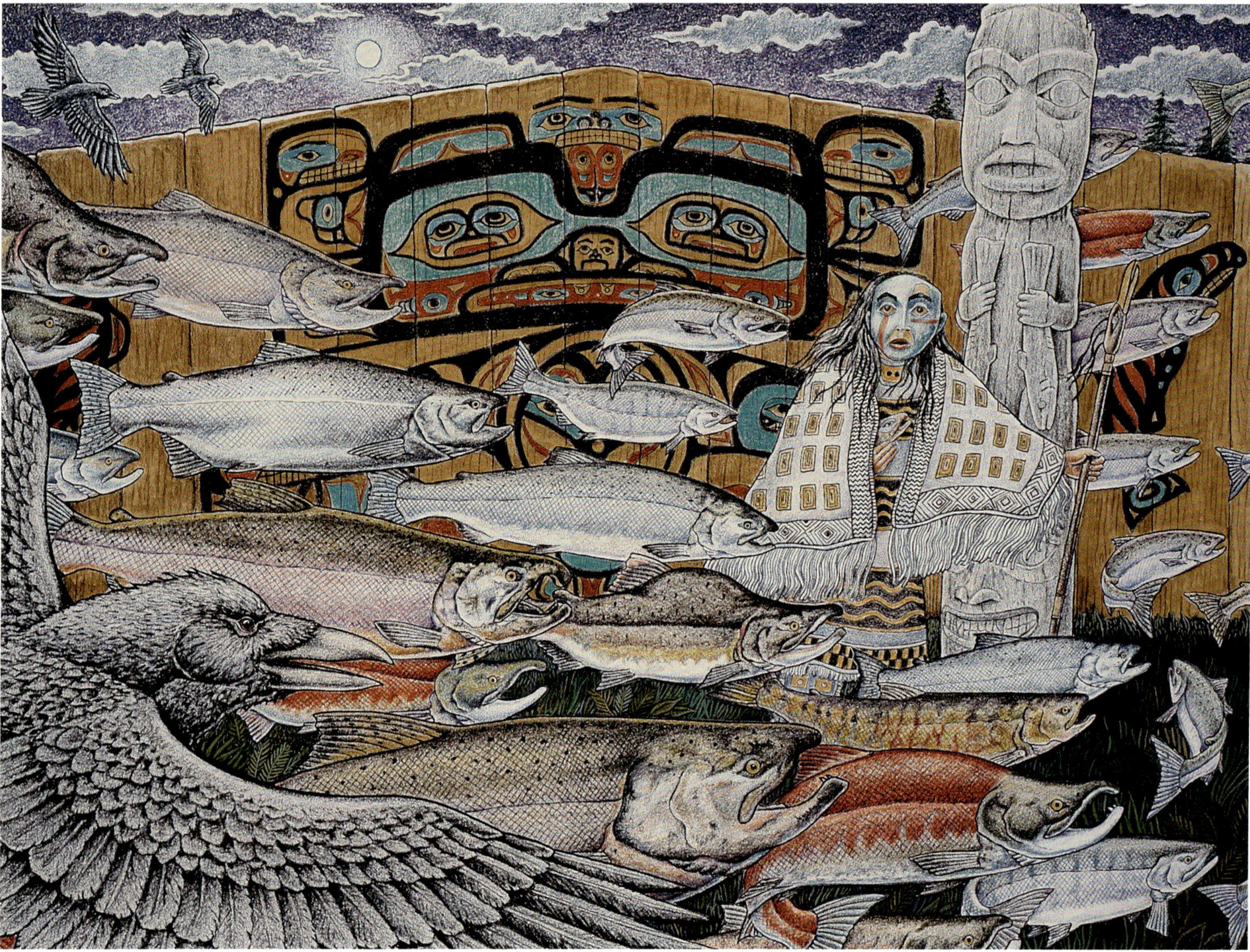

After moving to Alaska I developed a deep appreciation for the indigenous cultures of the Northwest Coast. I was surrounded by magnificent Tlingit, Haida, and Tsimshian carvings, and the stories and meanings of the art were a source of endless fascination for me. One of our local Tlingit tribes, the Tongass, has a wonderful story about a marriage between Fog Woman and Raven. Raven had only bony sculpins to eat before he met this mythical woman who lived at the head of Anan Creek, just north of Ketchikan. She gave him salmon, which she produced in a swirl of fog from the woven hat she wore on her head. They married, but Raven soon started to treat her badly, bragging to his friends about his tasty new fish and taking all the credit for their creation. One day when he struck her, she finally had enough of him and left, taking all of the salmon with her. In one version of the story I've heard, she allowed the salmon to come back once a year because she took pity on Raven as he cried out for forgiveness when she left. This drawing shows Fog Woman as she gathers the salmon around her in preparation to leave.

DANCE OF THE FISH CHARMERS

1985, colored pencil on paper, 18 in. × 22 in.

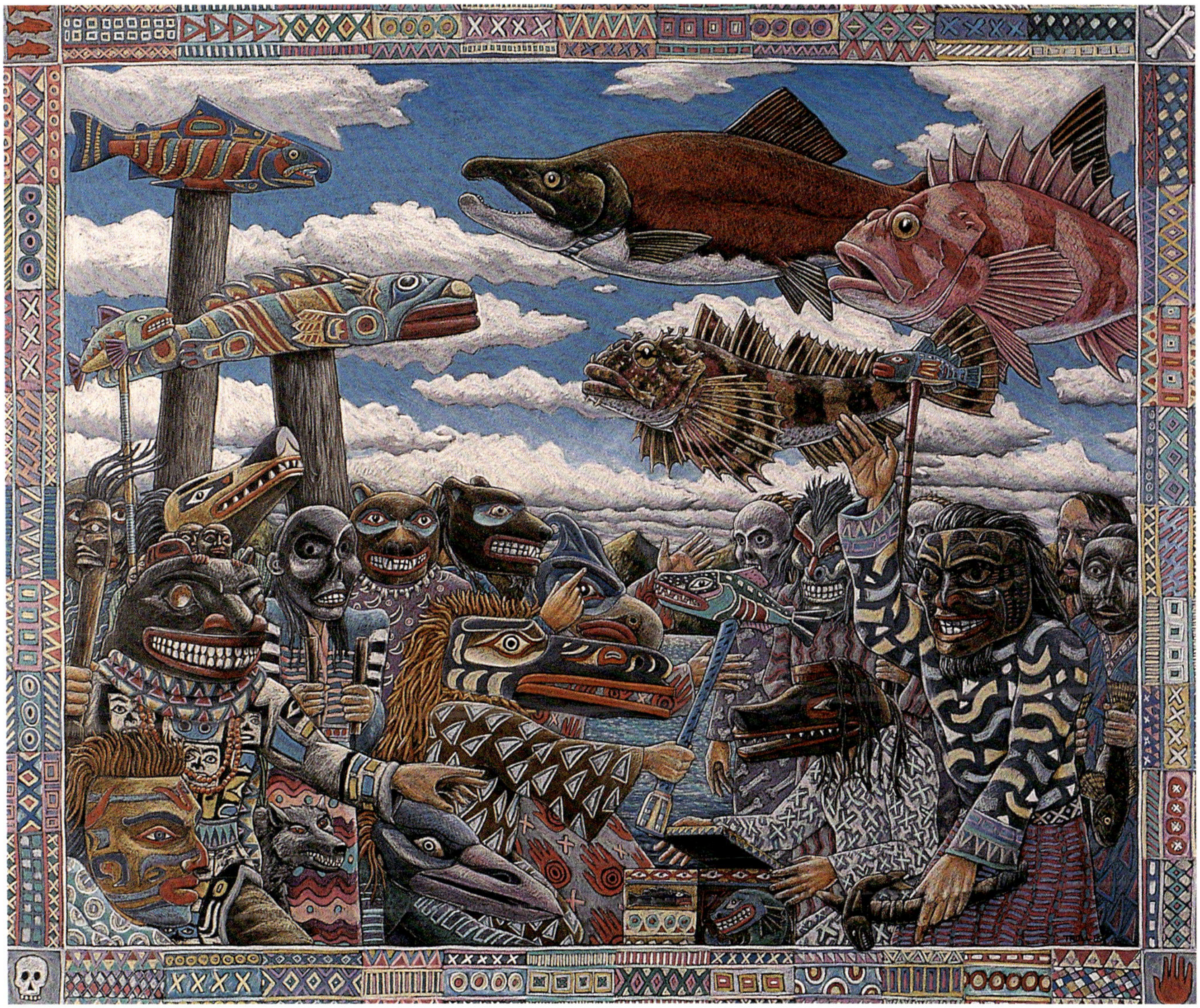

I IMAGINED two clans meeting somewhere along the coast with magical totemic fish floating in the sky. This drawing features a real cultural mix, with Northwest Coast art and elements from Tibetan-style fabrics stirred together with my own imagination. I can be seen at the far right in the background, appearing like Alfred Hitchcock in one of his films. I like the drama in this one. Even *I* want to know what's in that box at the center.

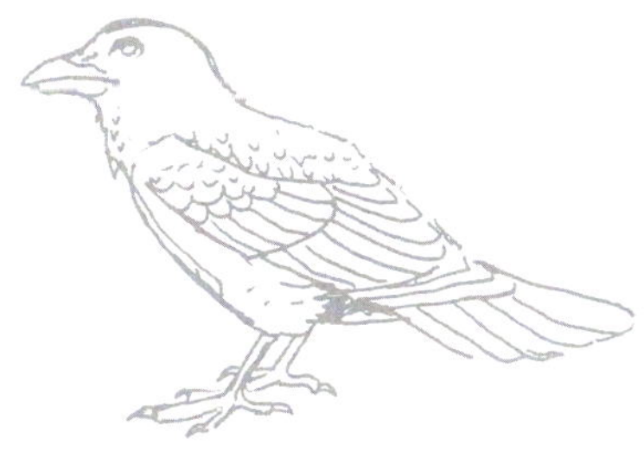

1985, mixed media on paper, 36 in. × 48 in.

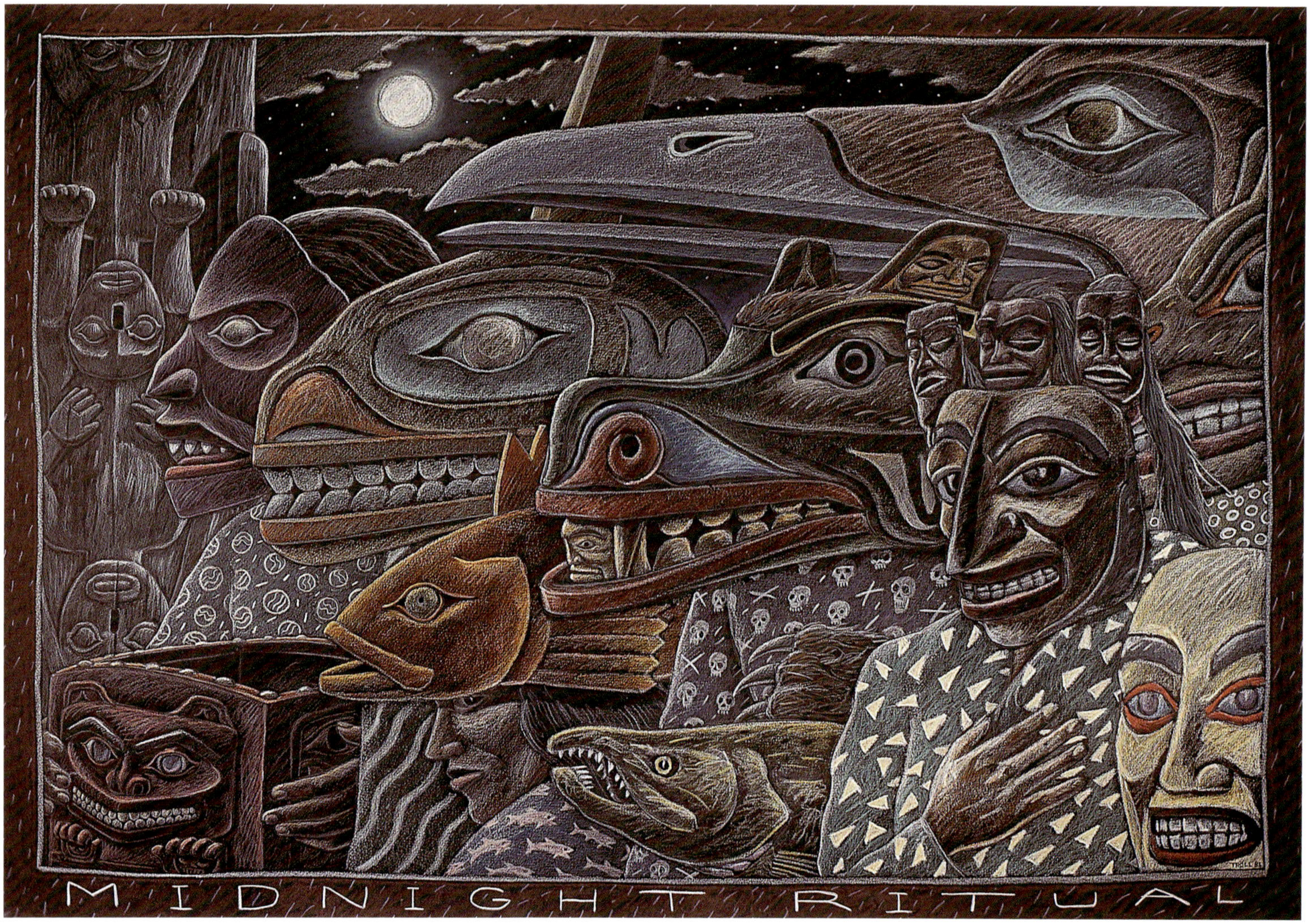

MIDNIGHT RITUAL

Masks, totems, and fabric patterns combine in a scene of late-night mystery. The totem pole on the left featuring rows of upside-down humans is in the totem park in Klawock, Alaska.

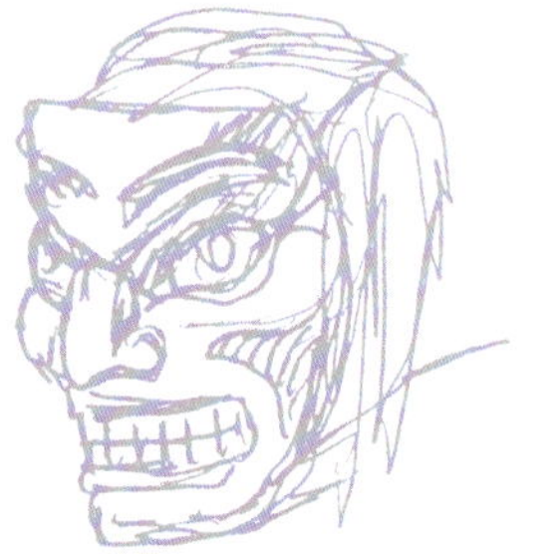

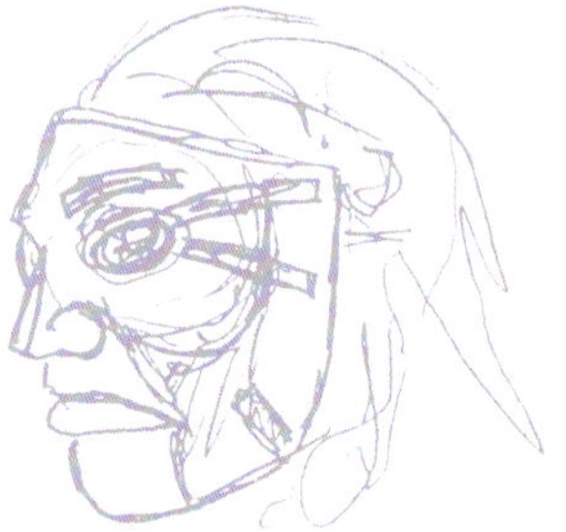

RAIN ON THE PARADE

1990, charcoal on paper, 60 in. × 78 in.

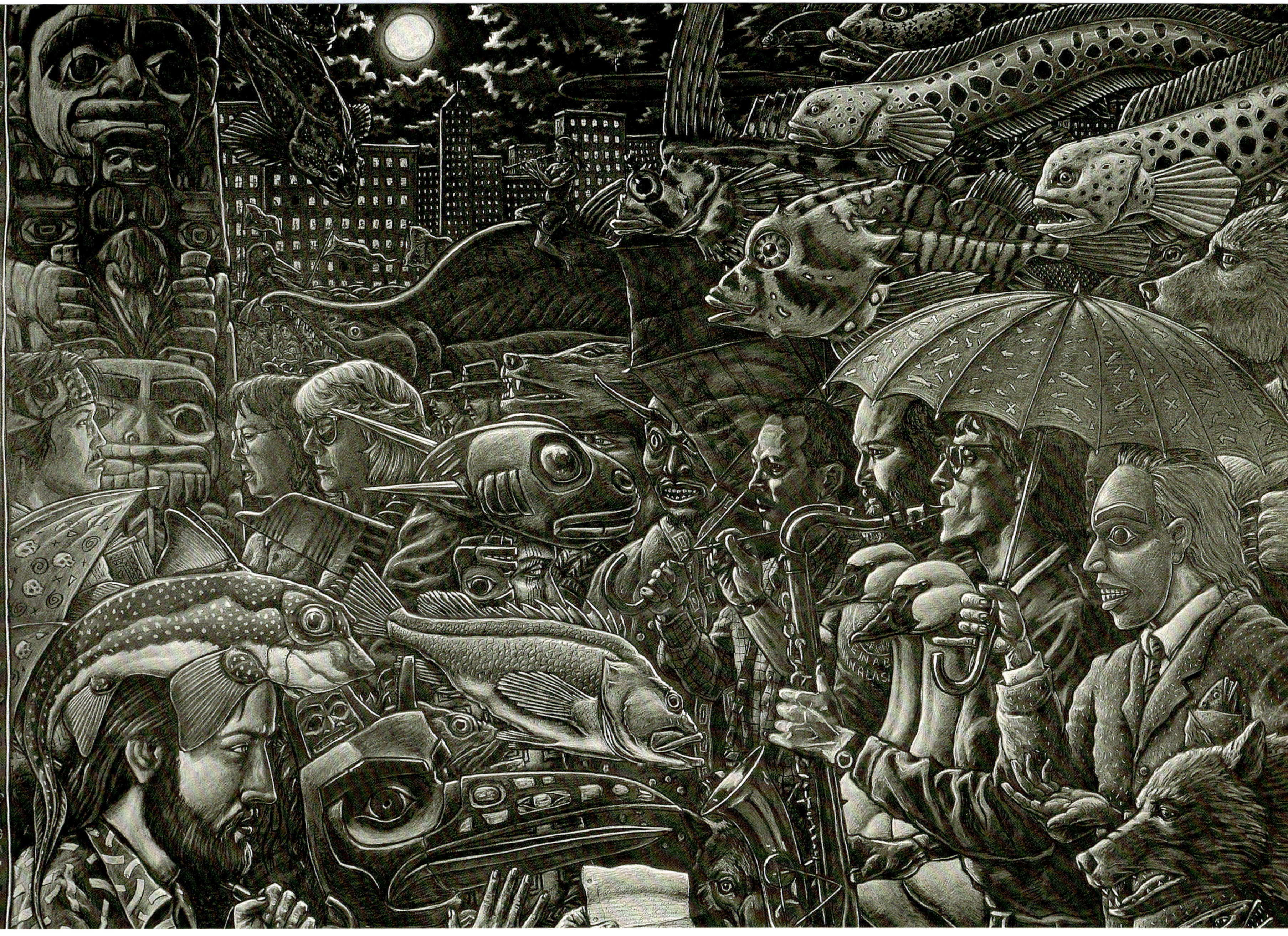

I've long been enthralled by the grand and bizarre opulence of parades. As a child I was hugely impressed by the local Mardi Gras in Mobile, Alabama. The Fourth of July parade in Ketchikan is a big deal, too, but it always misses the mark for me in representing what this strange and beautiful land called Alaska is really about. So I decided to do my own artist-biologist version of a "half small-town Ketchikan, half big-city Seattle"–style parade. Several of my artist and musician pals can be seen in this gigantic charcoal dust drawing, along with many of my favorite fish (wolf eels, a sailfin sculpin, a grunt sculpin, and a musically inclined red snapper). I'm in the lower left with a ratfish on my head. Farther back in the crowd is a Tlingit sculpin headdress as a kind of parallel presence. My wife, Michelle, and her good friend Hilary are in the drawing as well. I rubbed black charcoal dust all over a gigantic piece of white paper, then drew the entire image with a kneaded eraser. My studio became a complete sooty mess and my lungs ached for weeks.

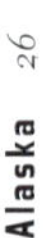

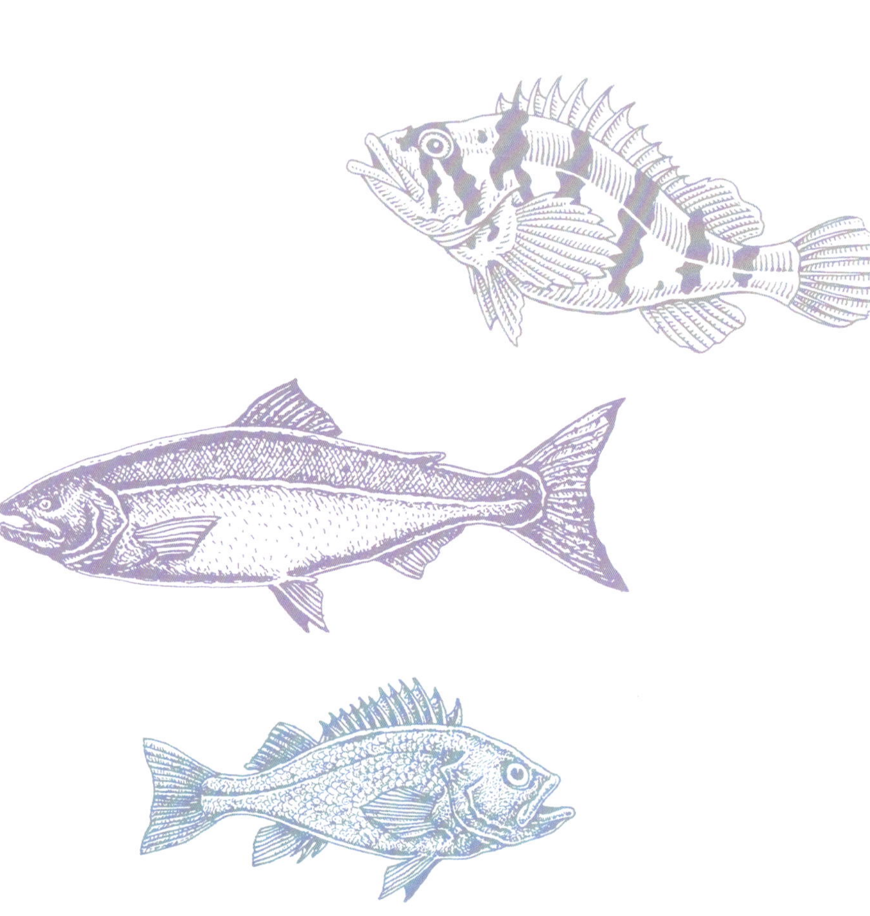

A VIEW BELOW the ocean in the waters near Kodiak, Alaska, depicting sea creatures of myriad aspects and types. I photographed fishing vessels on the grid north of town for this piece and gave the boats names that are significant to me: the *Michelle K* (my wife), the *True Blue* (my dream boat: a double-ended wooden troller), and the *Mary Lou* (one of the songs my art band Zuzu and the Robot Slave Boys played during my graduate school days).

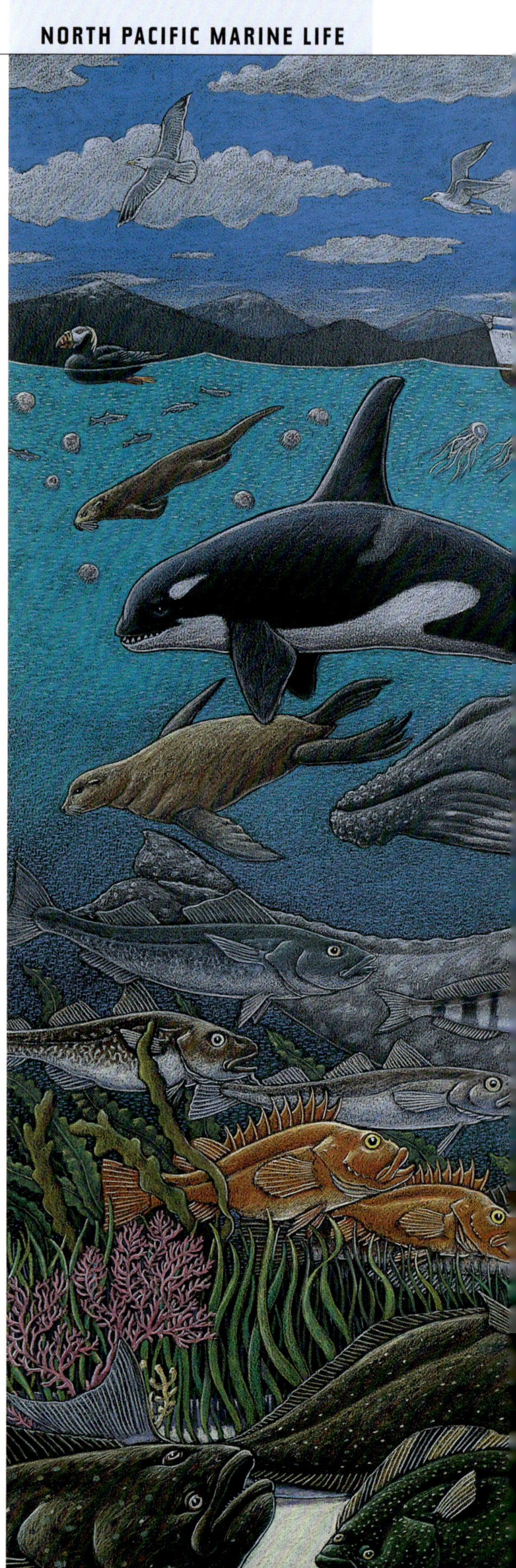

1986, colored pencil on paper, 22 in. × 30 in.

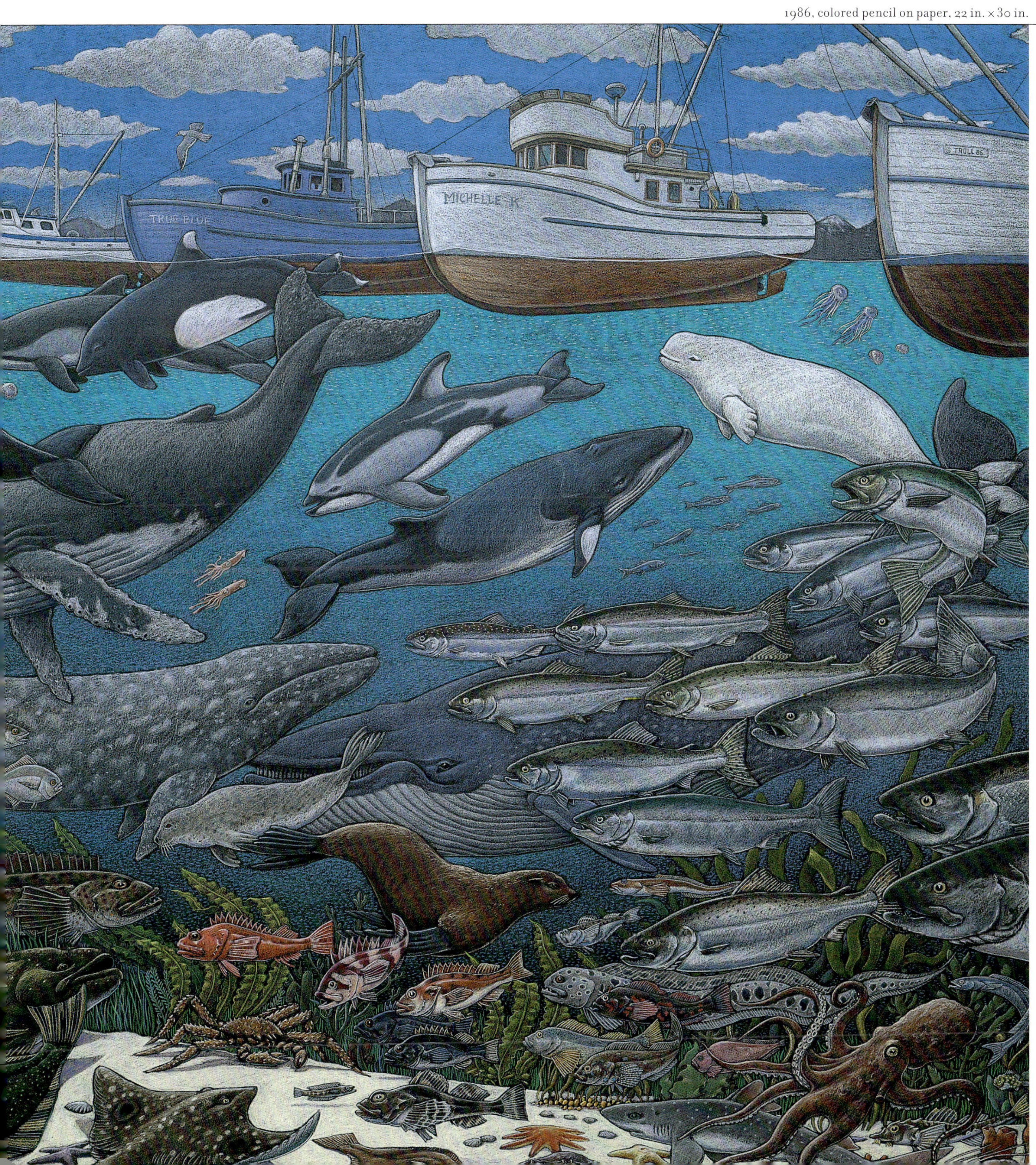

2000, pen, ink, and watercolor on paper, 13 in. × 13 in.

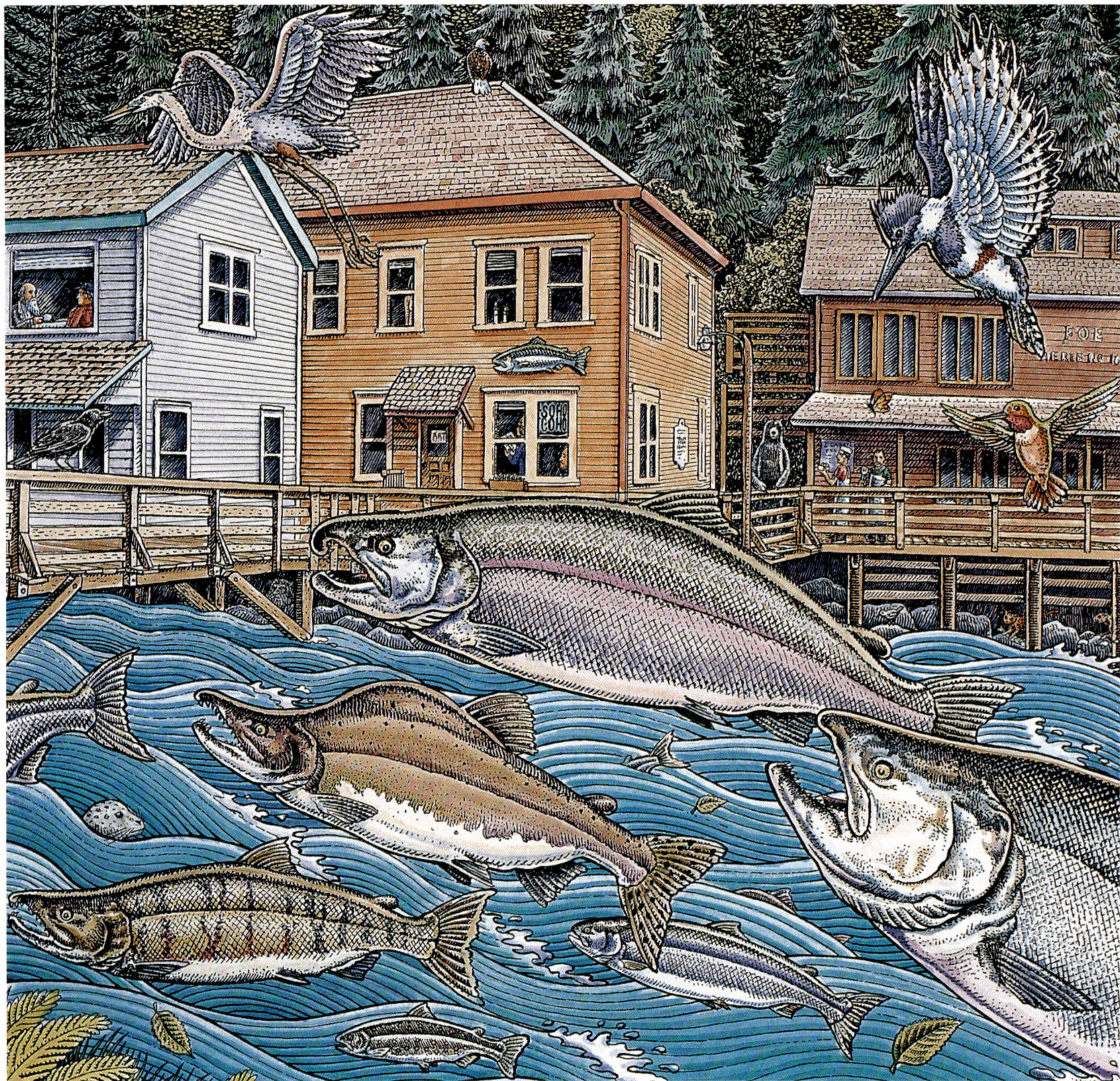

Our gallery, the Soho Coho, The Eagles Club, and the Good Fortune Chinese restaurant are shown in this drawing, along with four species of salmon, a steelhead trout, a kingfisher, and a great blue heron. You can spot Charles Darwin enjoying lunch at the restaurant. The buildings are located on Creek Street in Ketchikan, a former red-light district that was closed down in 1954, the year I was born. Most of the buildings on the street are original structures. Our gallery was an old dance hall, with rooms upstairs available by the hour. The standing joke is that Creek Street was "a place where both men and salmon headed upstream to spawn."

1987, charcoal on paper, 40 in. × 26 in.

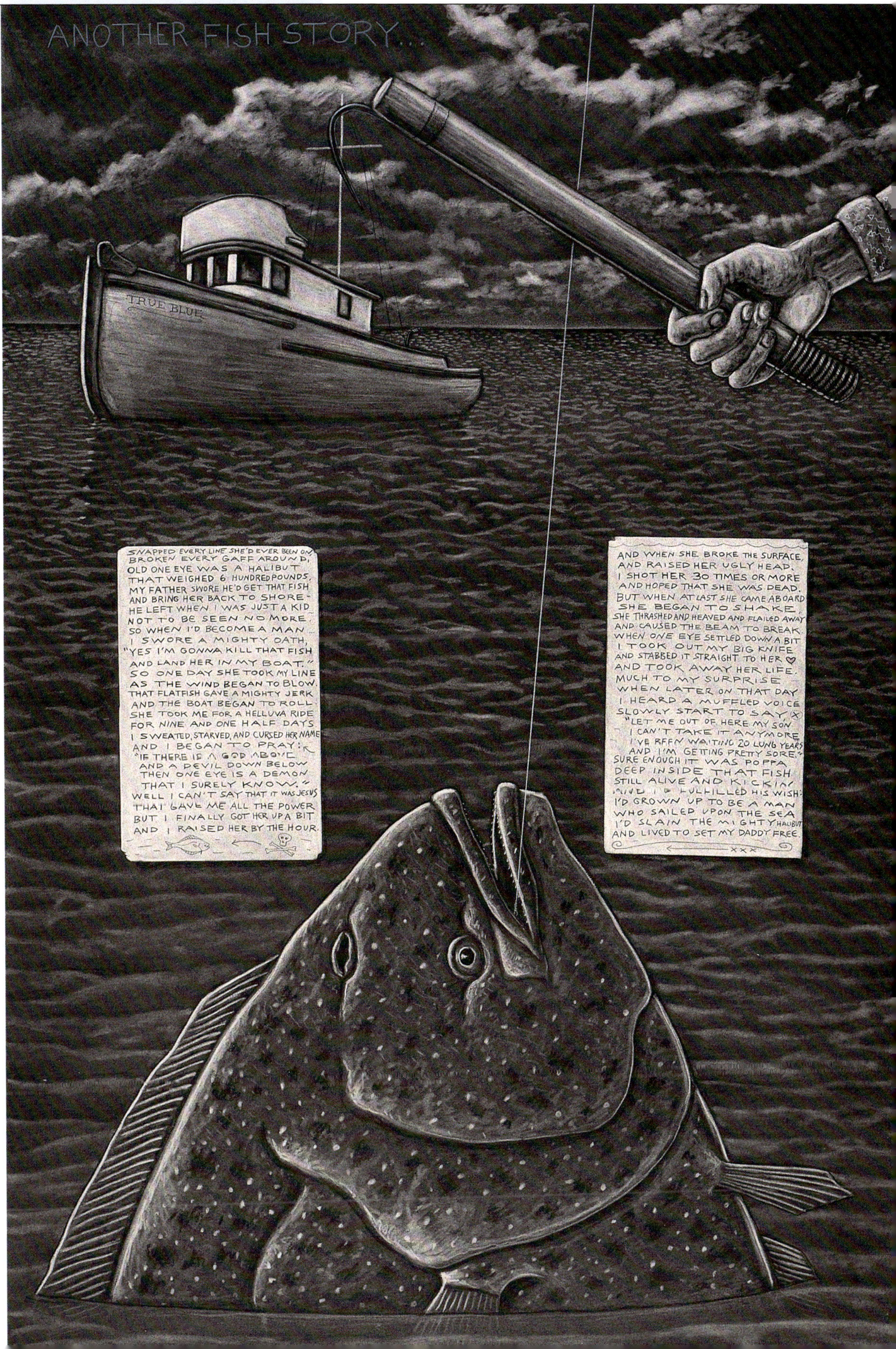

I WAS IN A GARAGE BAND called the Squawking Fish for a while in the late eighties. We wrote most of our own tunes and actually got paid a few bucks to perform at a few functions around town. I wrote a goofy country-western song for the band called "Old One Eye" but never could figure out the right guitar chords for it. In the end I figured it would make a more interesting drawing.

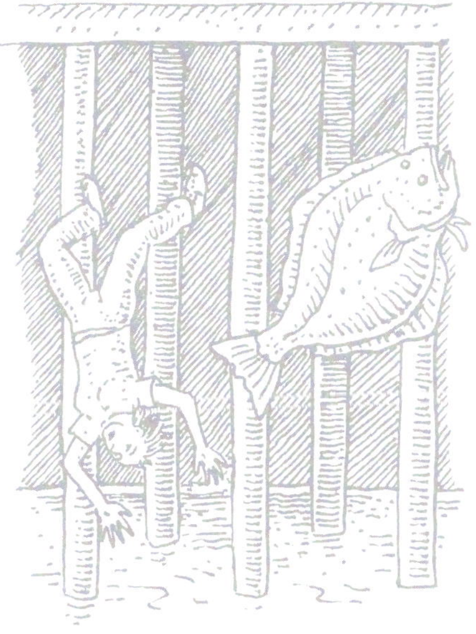

1983, colored pencil and spray paint on paper, 24 in. × 36 in.

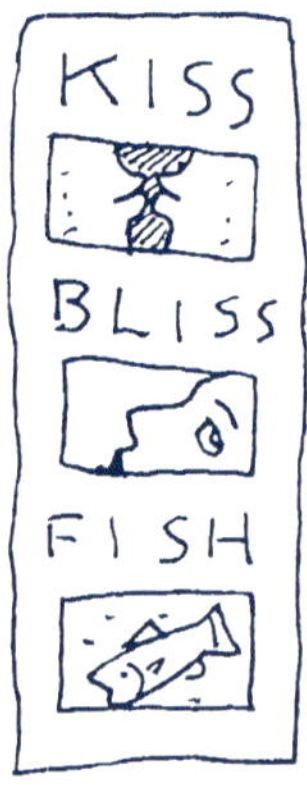

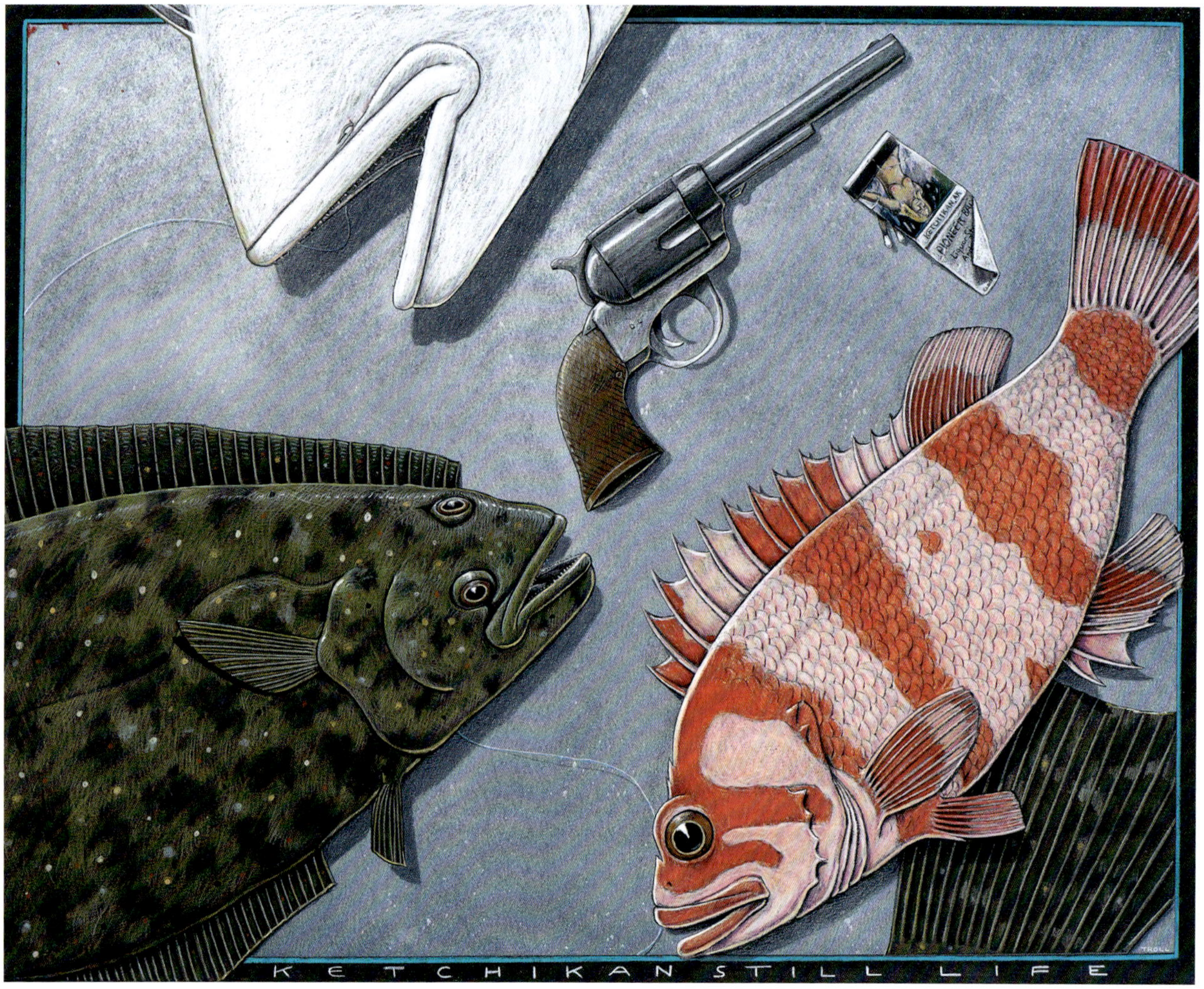

FISH, GUNS, AND BARS, three things that make Ketchikan what it is.

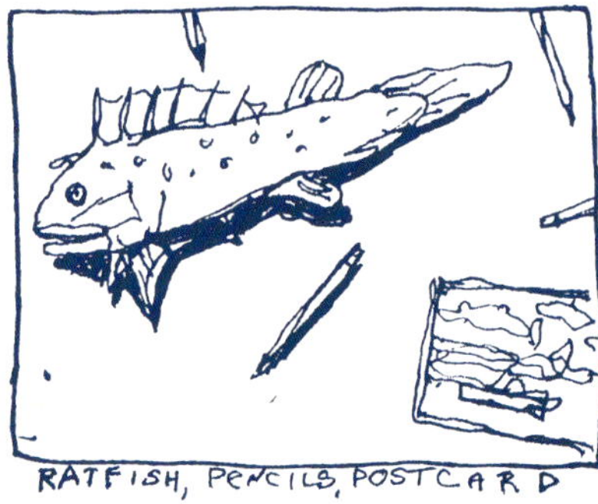

1991, pastel on paper, 22 in. × 30 in.

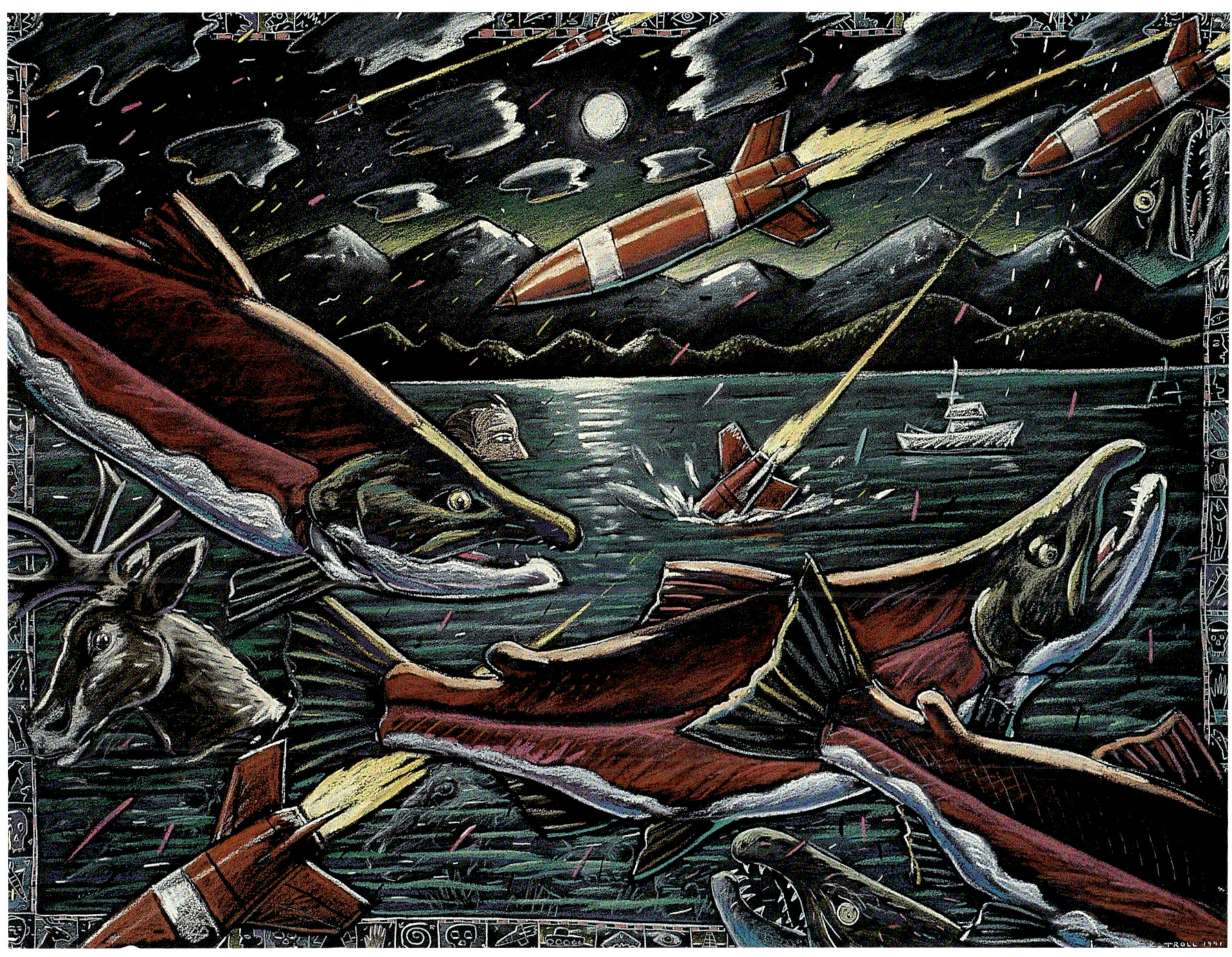

THIS IS a fairly loose pastel drawing I did shortly after the first Gulf War, hence the sleek-looking Scud missiles. Every time the Middle East oil supply is interrupted, the political rhetoric heats up about opening Alaska's North Slope to further oil exploration. So in a roundabout way, missiles fired from Baghdad end up landing right here in our own neck of the woods.

1986, alkyd and oil on canvas, 60 in. × 144 in.

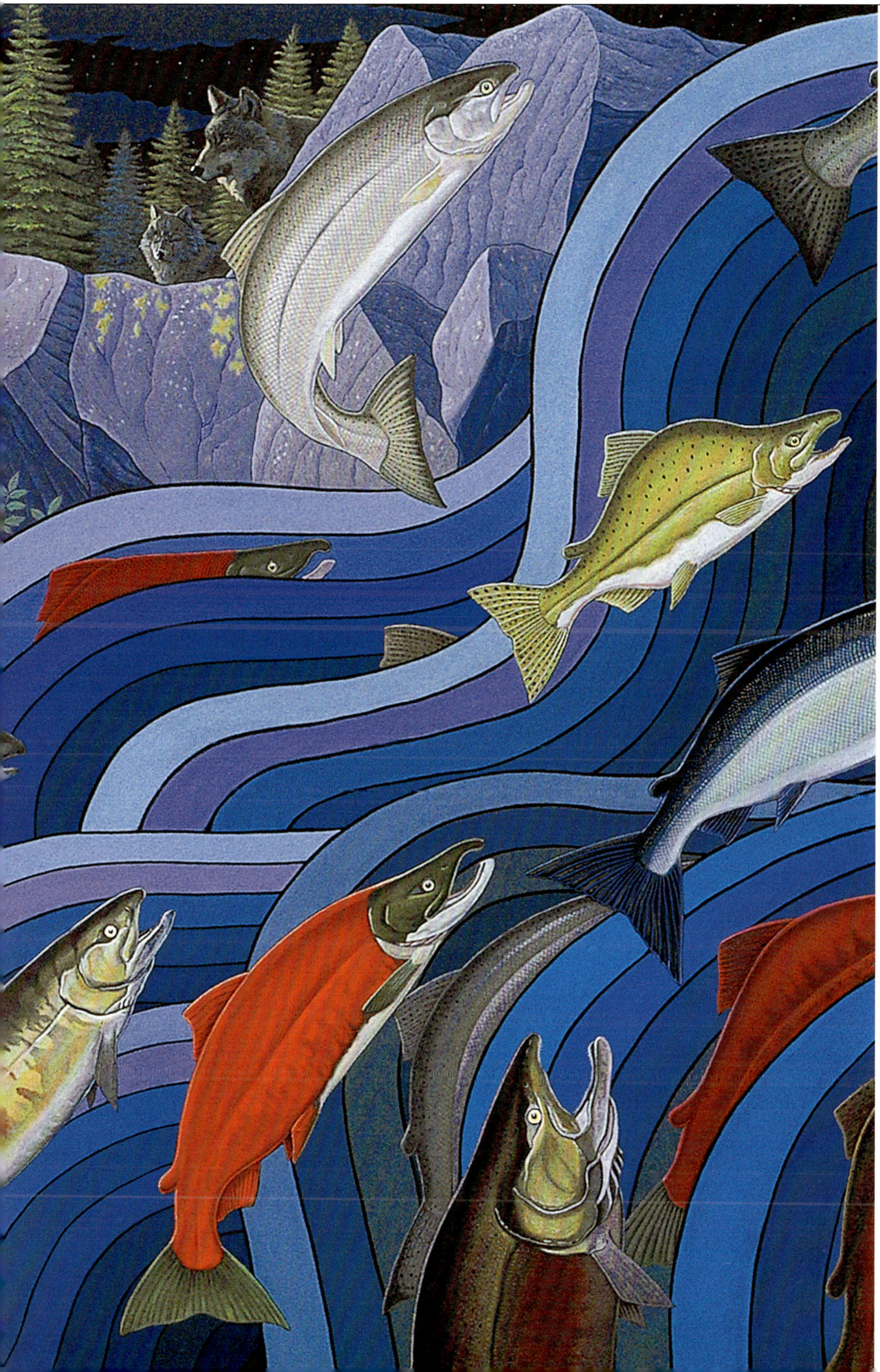

THIS WAS one of my first major commissions, done for a local elementary school. I still remember the day I got the phone call from the local committee member saying I had gotten the job. I was ecstatic because it meant I could finally quit my day job and concentrate on making art. I spent the next few months on this mural, painting all the Alaskan salmon species, steelhead, Dolly Varden, a lone black bear, and a couple of wolves against big undulating neo-Japanese–style water patterns. To help me in this, I wandered down the block and spent many days photographing freshly caught salmon at the local fish processors.

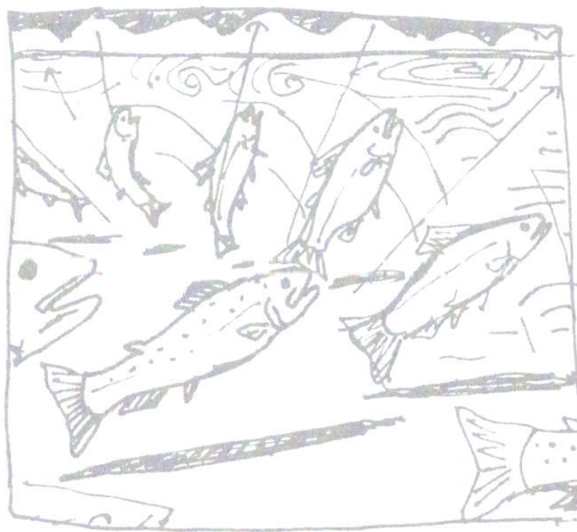

DYNAMICS OF A SALMON FLIP

1989, pen, ink, and watercolor on paper, 7.5 in. × 8 in.

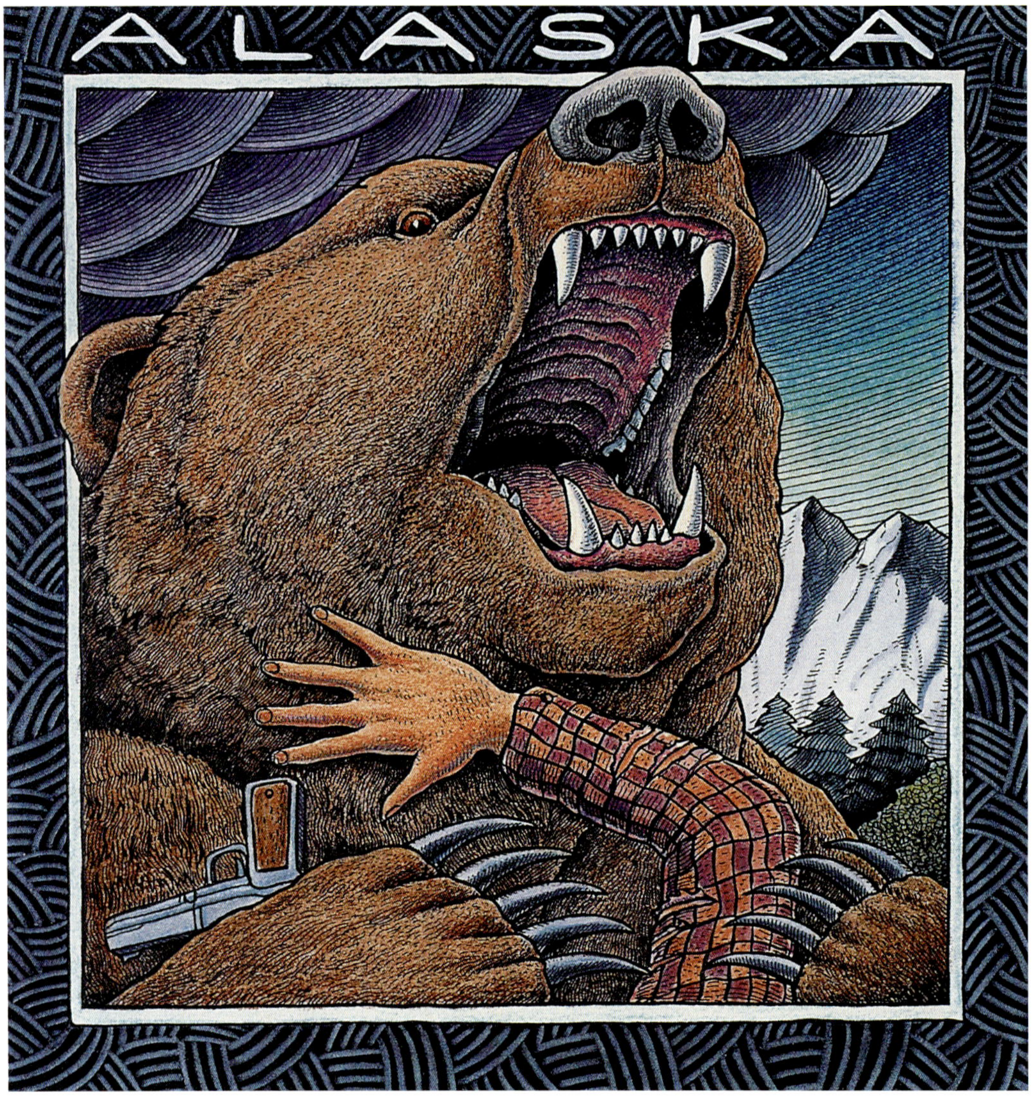

Alaskans love to scare the hell out of each other by telling harrowing and gruesome bear stories around campfires late at night, and I'm no exception. I think the fascination fills some deep biological need to be keenly aware of animals that can actually eat us. As a result, I am now more than a little "bearanoid" when I head off into the deep forests that surround our town. This drawing was done shortly after a particularly horrible incident on nearby Baranof Island in which a hapless deer hunter was eaten by a large brown bear *(Ursus horribilis!)*. When the man was found the next day, all that remained of him was his spinal column. Mental images like that tend to stick with you, and I dwelled on the incident for days before I drew this ditty. When I released it as a T-shirt, though, I made the title a little less subtle: "Snacktime."

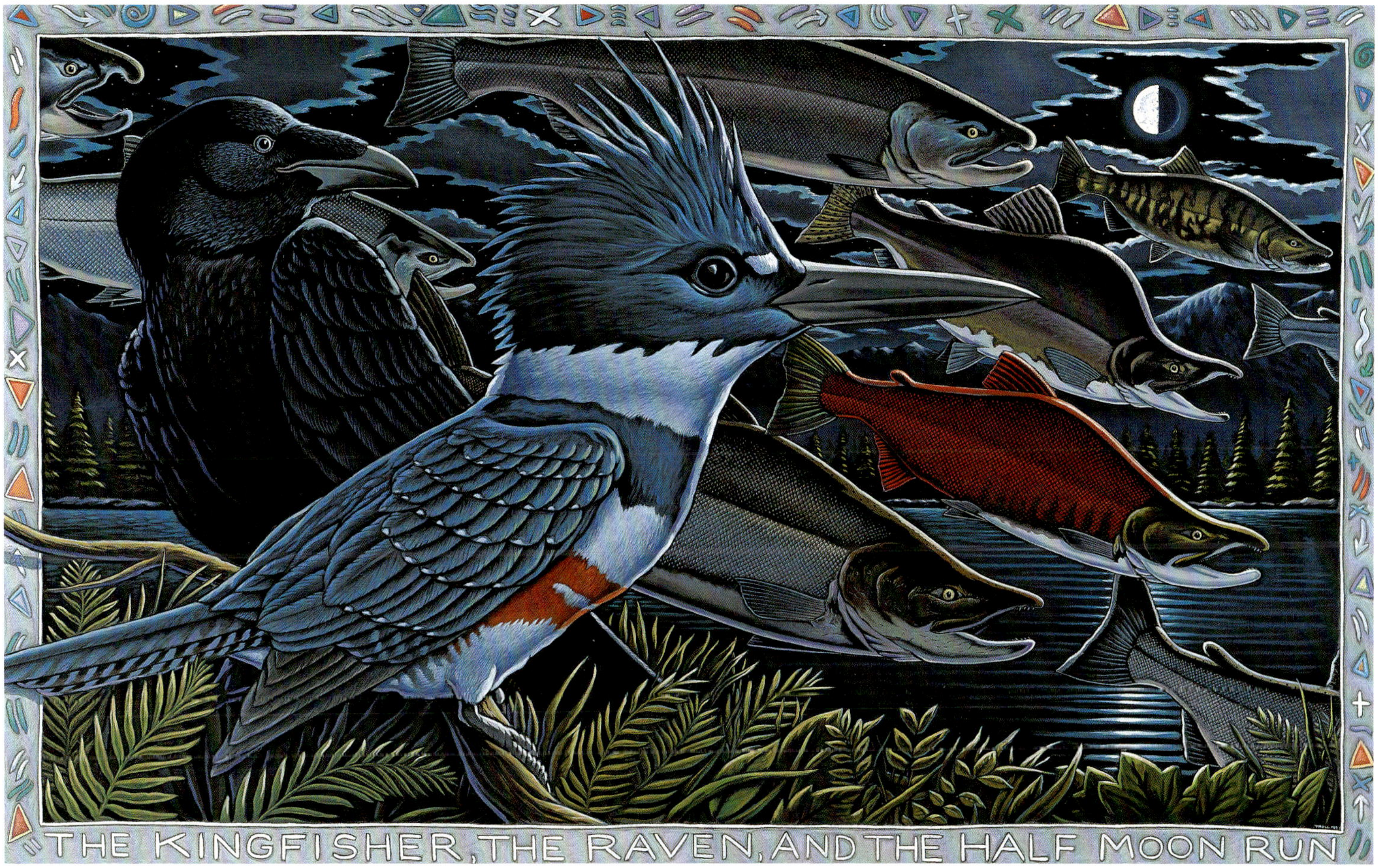

1989, oil on canvas, 48 in. × 78 in.

IT'S NOT UNCOMMON for me to have dreams of fish floating in the sky. I wondered if a bird like the belted kingfisher, which spends nearly all its waking energy tracking the movements of fish, spends its nights dreaming of them as well. Raven, ever the clever opportunist, looks on over Kingfisher's shoulder.

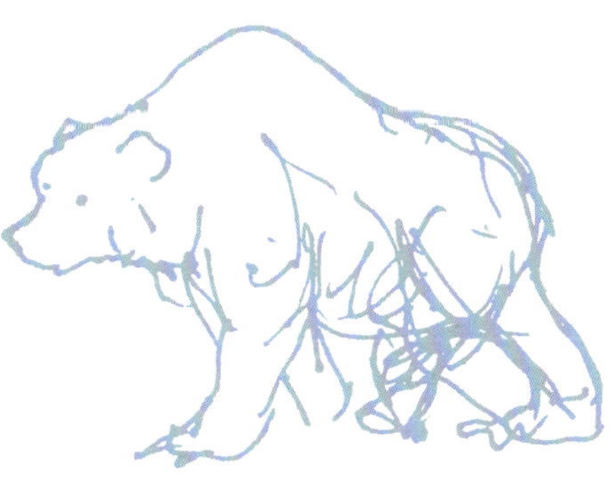

1988, colored pencil on paper, 32 in. × 40 in.

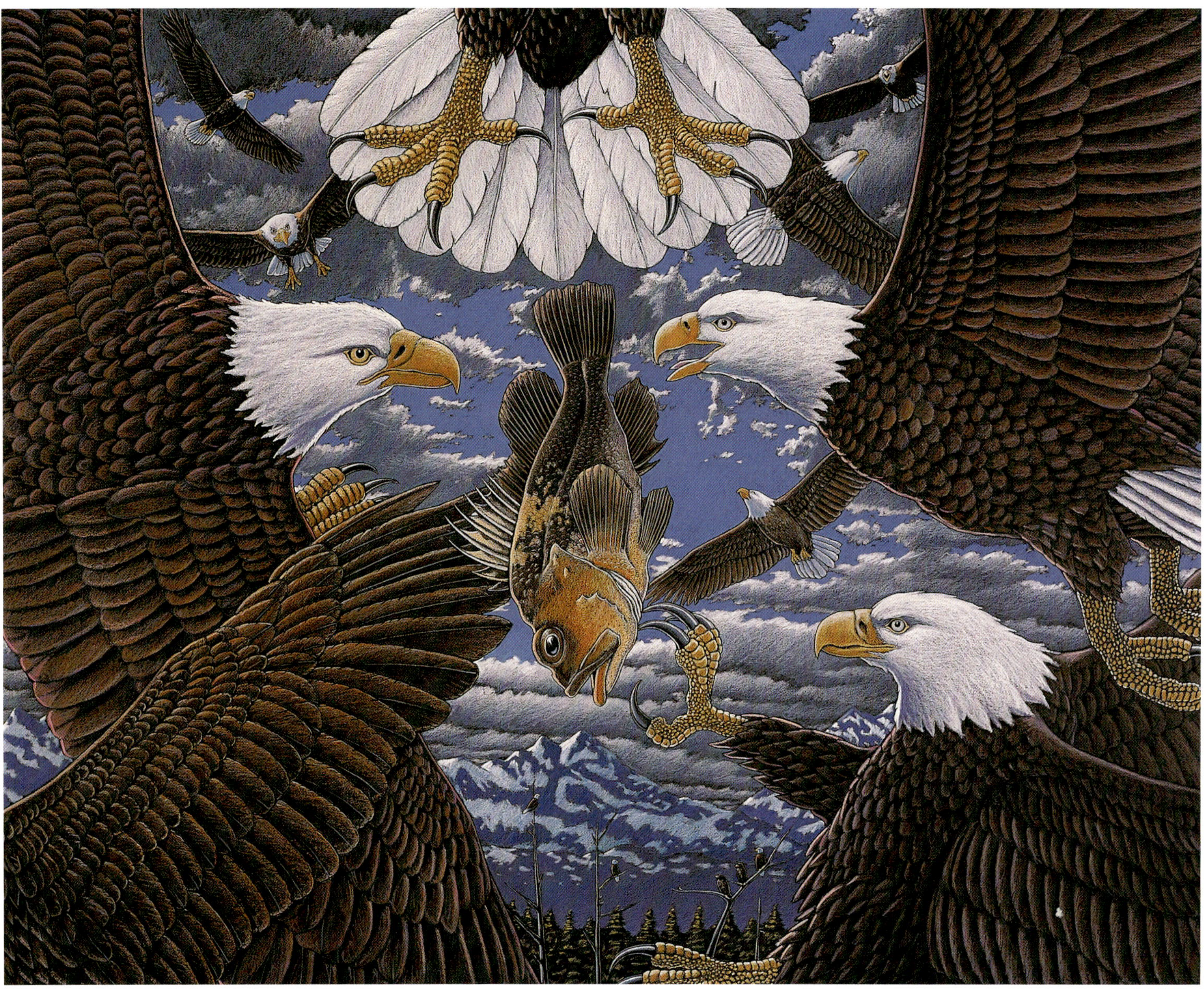

SPORT FISHERMEN in Alaska typically target more glamorous species like salmon and halibut, referring to the smaller rockfish they occasionally hook as "eagle bait." Bald eagles often follow the sportfishing boats, and when a rockfish is tossed back into the water they'll swoop down and snag it from the surface. That fishermen would treat rockfish so blithely has always puzzled me. In the lower forty-eight states smallmouth bass fishermen would love to catch a fish the size of an Alaskan rockfish. Go figure. Rockfish are also tasty as all get-out.

 1988, oil pastel on paper, 40 in. × 32 in.

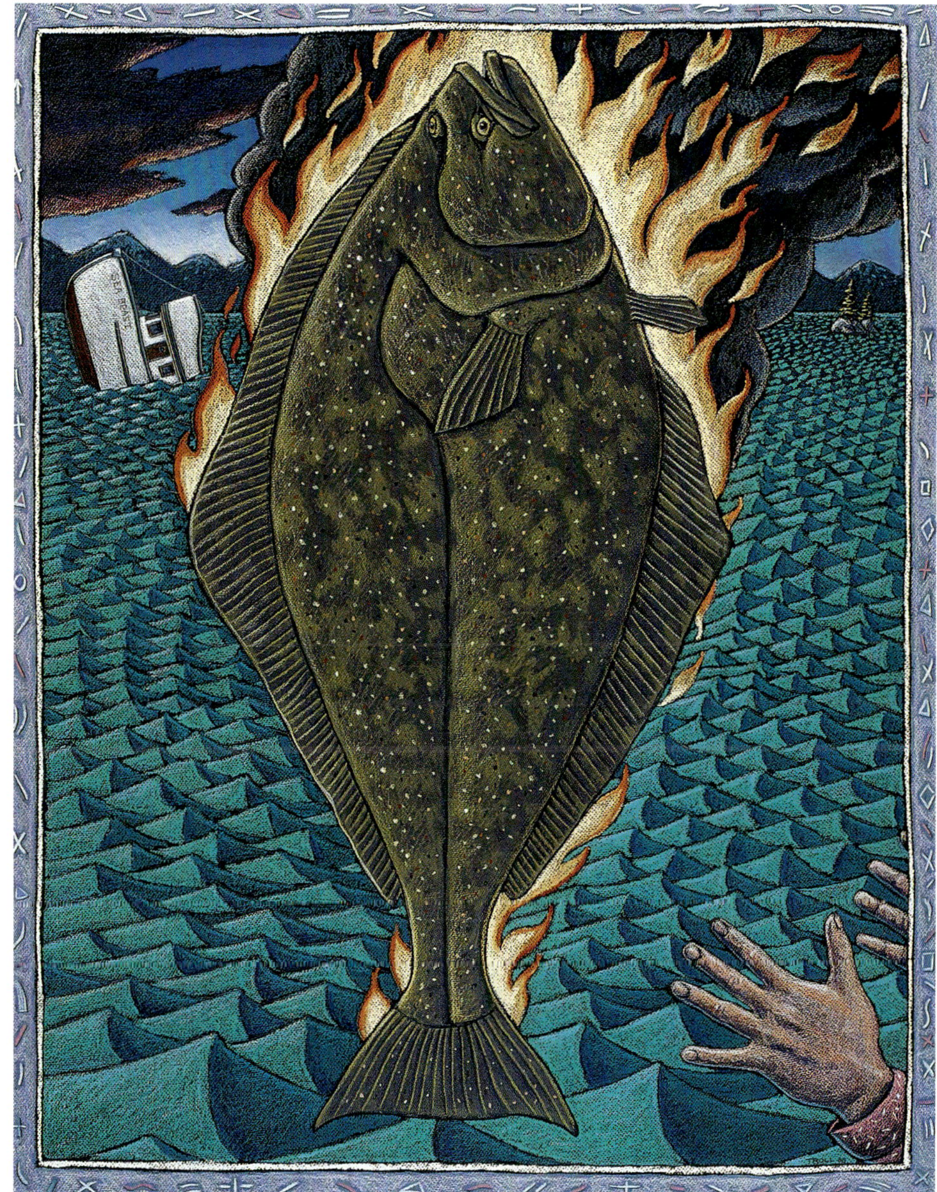

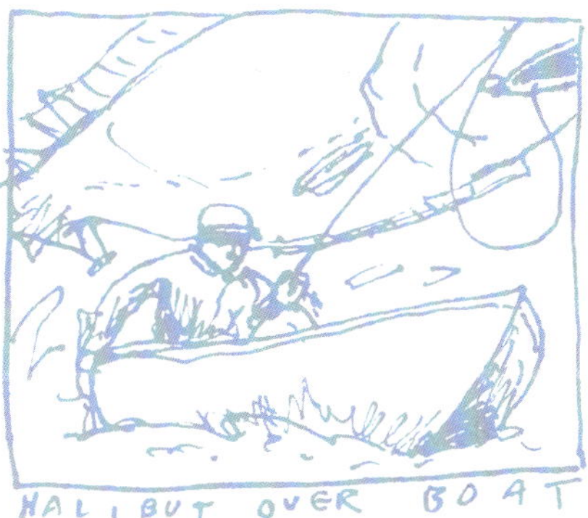

SOON AFTER I moved to Alaska in 1983 I started fishing on the waters of the Inside Passage in my brother-in-law's sixteen-foot Lund skiff and caught quite a few halibut, which were the stars of many delicious meals. Fishermen fish and fishermen lie, and I heard astounding yarns around the docks of halibut reaching monstrous proportions and of heroic struggles with these flat leviathans. I heard about three-hundred-pound halibut breaking legs or maiming fishermen as the great fish thrashed on deck. Serious fishermen pack pistols on their boats to subdue them. One account told of a halibut that actually killed a guy named Joe Cash—a story that Brad and I put in our book *Shocking Fish Tales.*

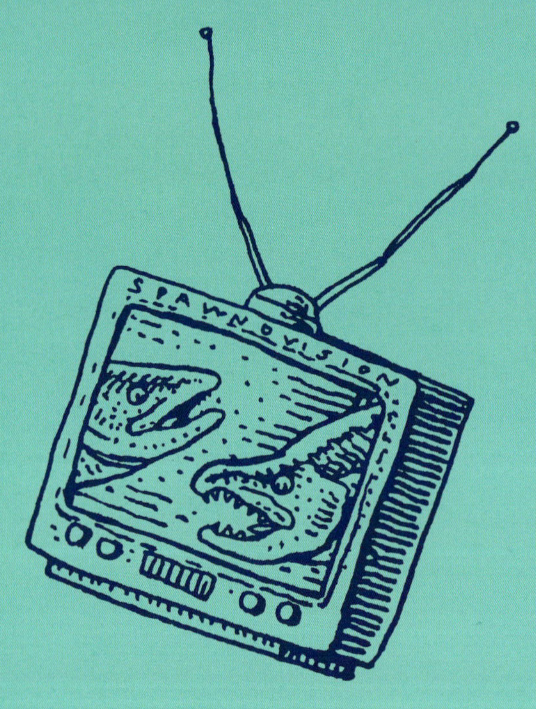
SPAWNOVISION

STREAM OF CONSCIOUSNESS

KNOW YOUR SALMON, BUB

1986, pastel on paper, 36 in. × 48 in.

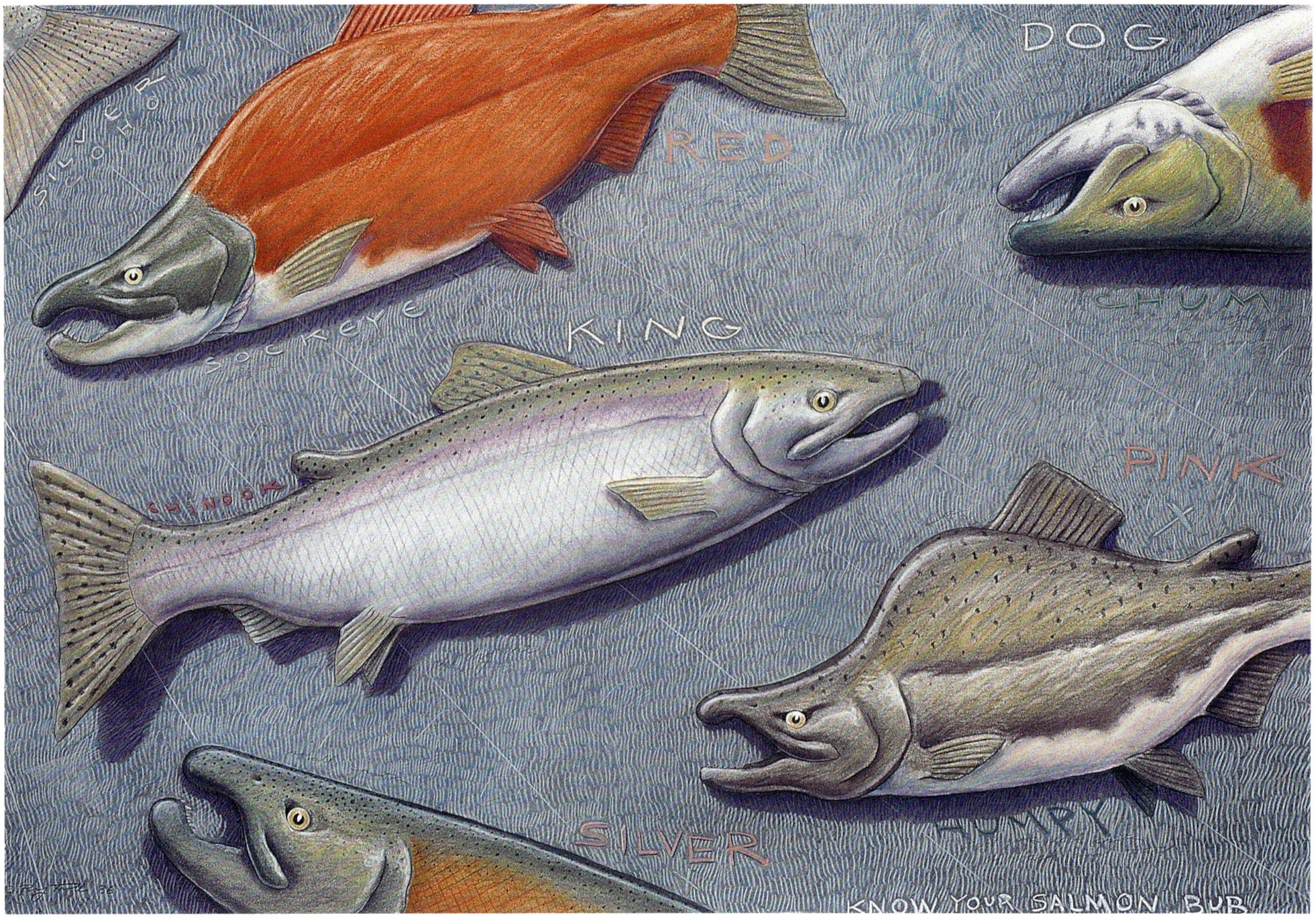

A PASTEL DRAWING depicting the splendid diversity of the genus *Oncorhynchus.*

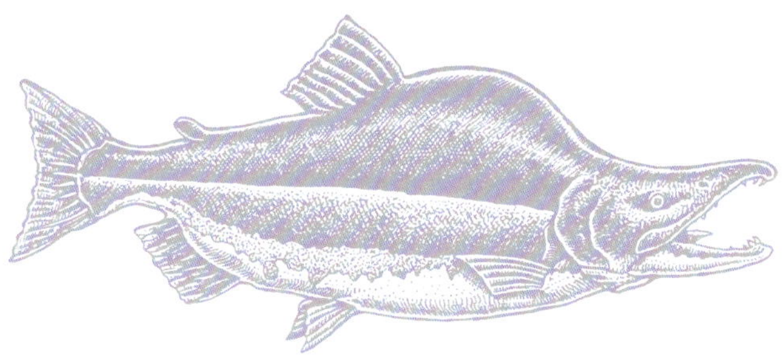

FISHSCAPE

1985, acrylic on canvas, 24 in. × 36 in.

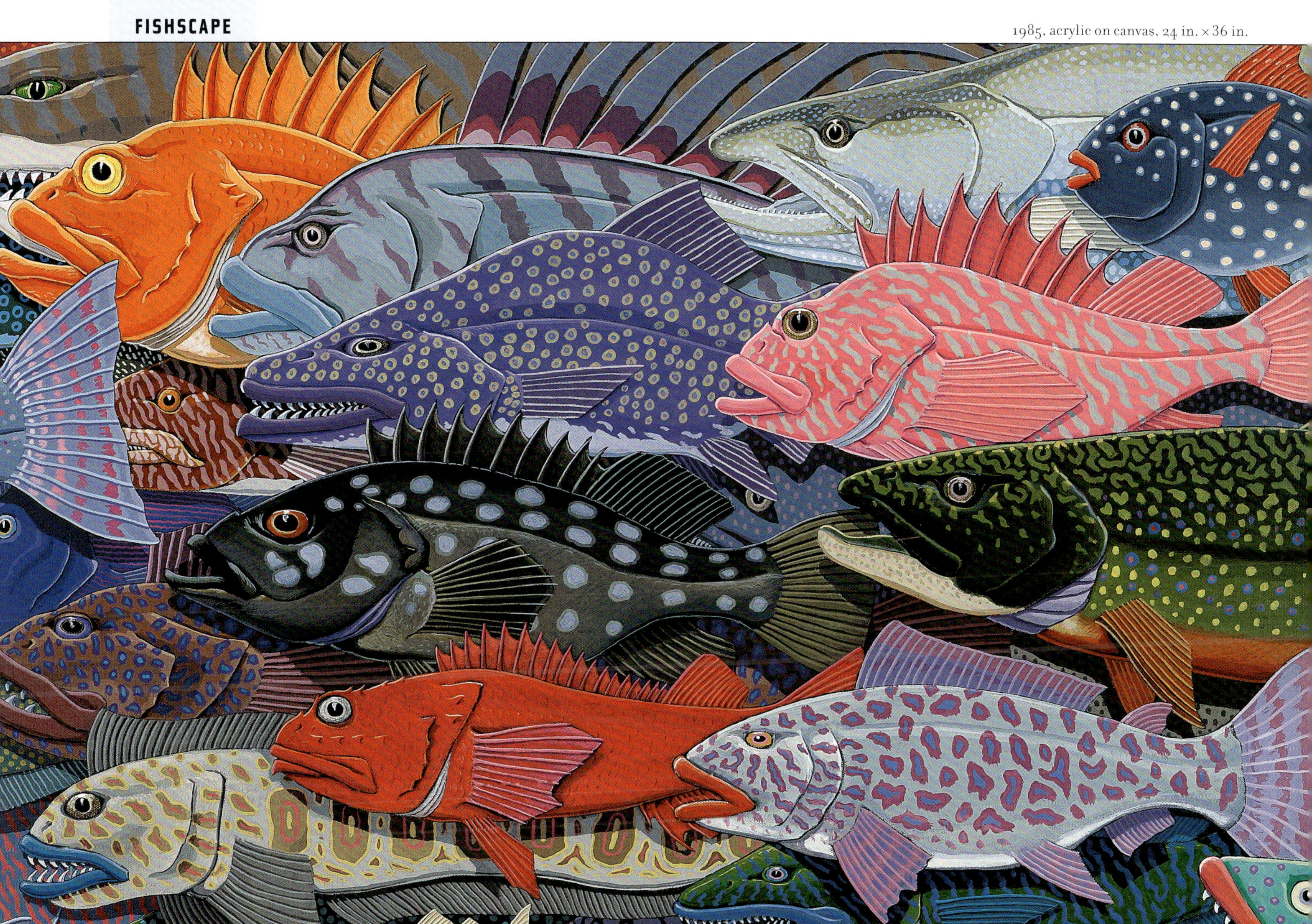

THIS IS an early painting with a mixture of real, kinda real, and totally unreal fish. Color and pattern were the overriding concerns in this composition.

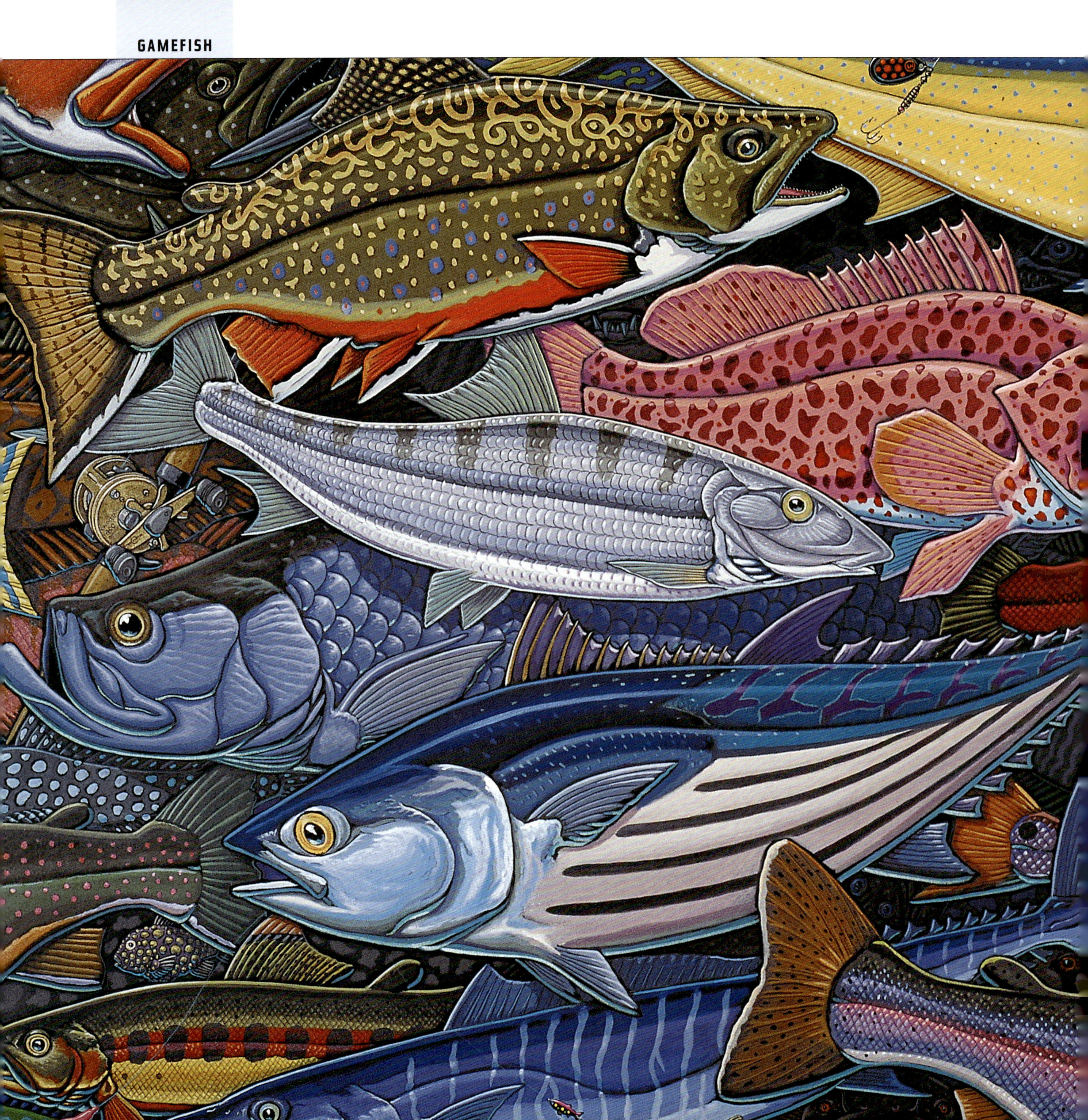

1996, acrylic on canvas, 60 in. × 84 in.

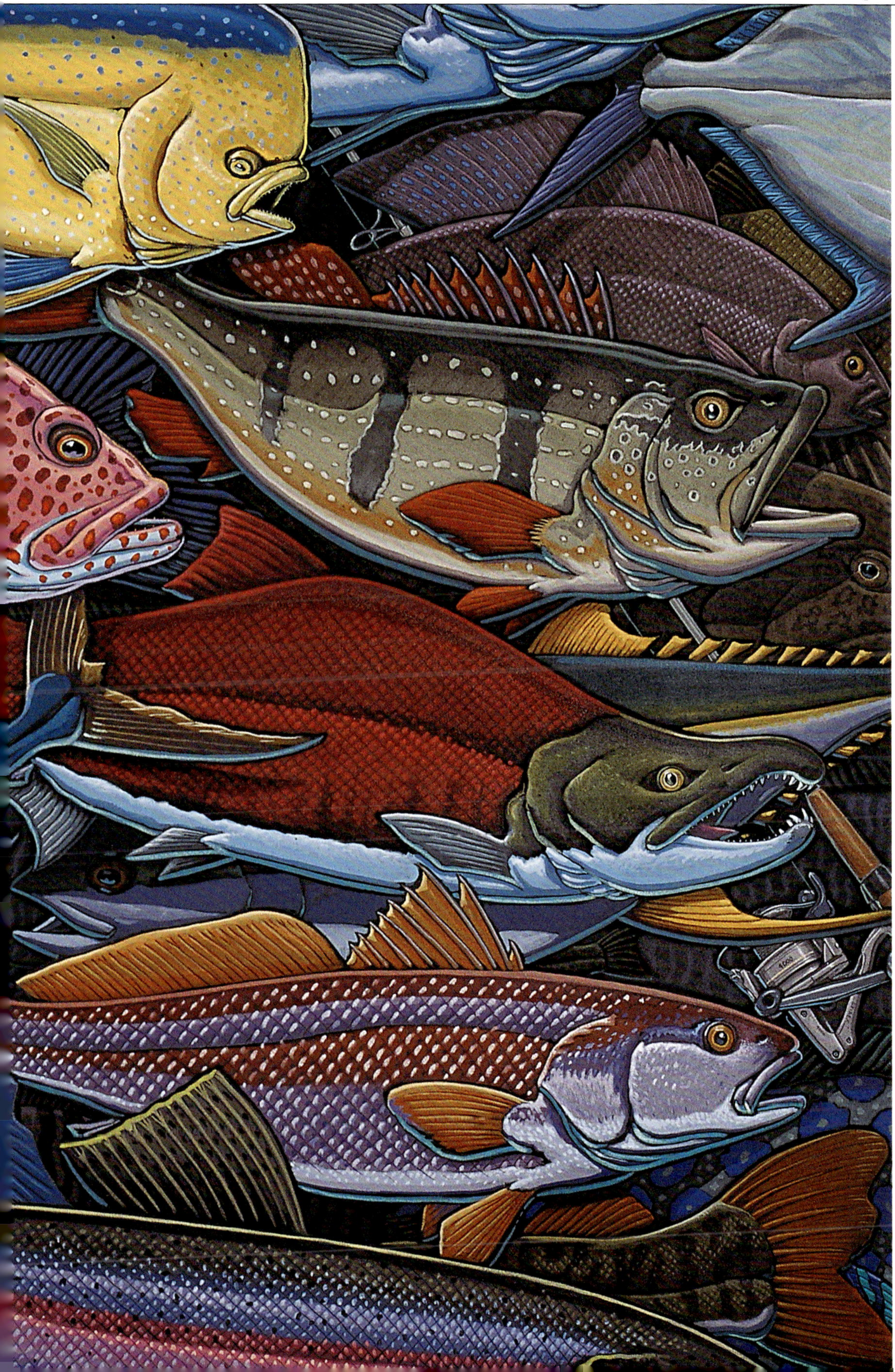

"Gamefish" are fish that are revered as suitable adversaries for the challenge-minded sportsperson, and in my typical fill-every-inch-of-available-space mode I've plugged the canvas with a colorful mélange of some of the more interesting ones. This acrylic painting was commissioned by the Shimano Corporation, a giant Japanese company that manufactures fishing gear. They used it as a catalog cover; if you look closely you can spot a few of their rods and reels in the background.

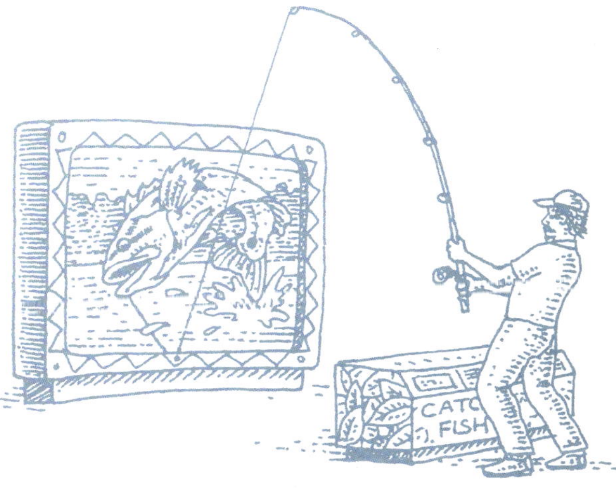

BOTTOM FISH OF THE NORTH PACIFIC

1983, colored pencil on paper, 28 in. × 22 in.

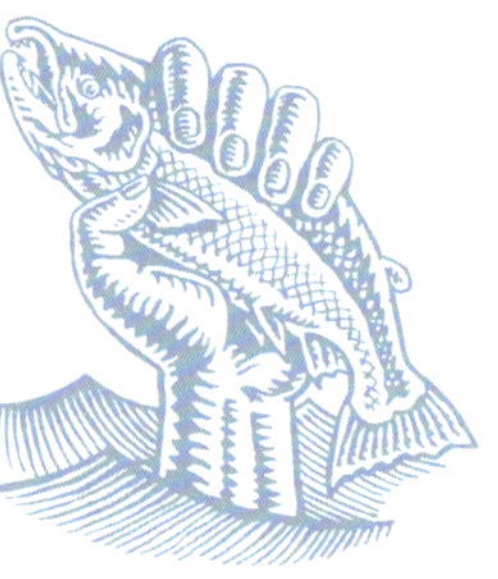

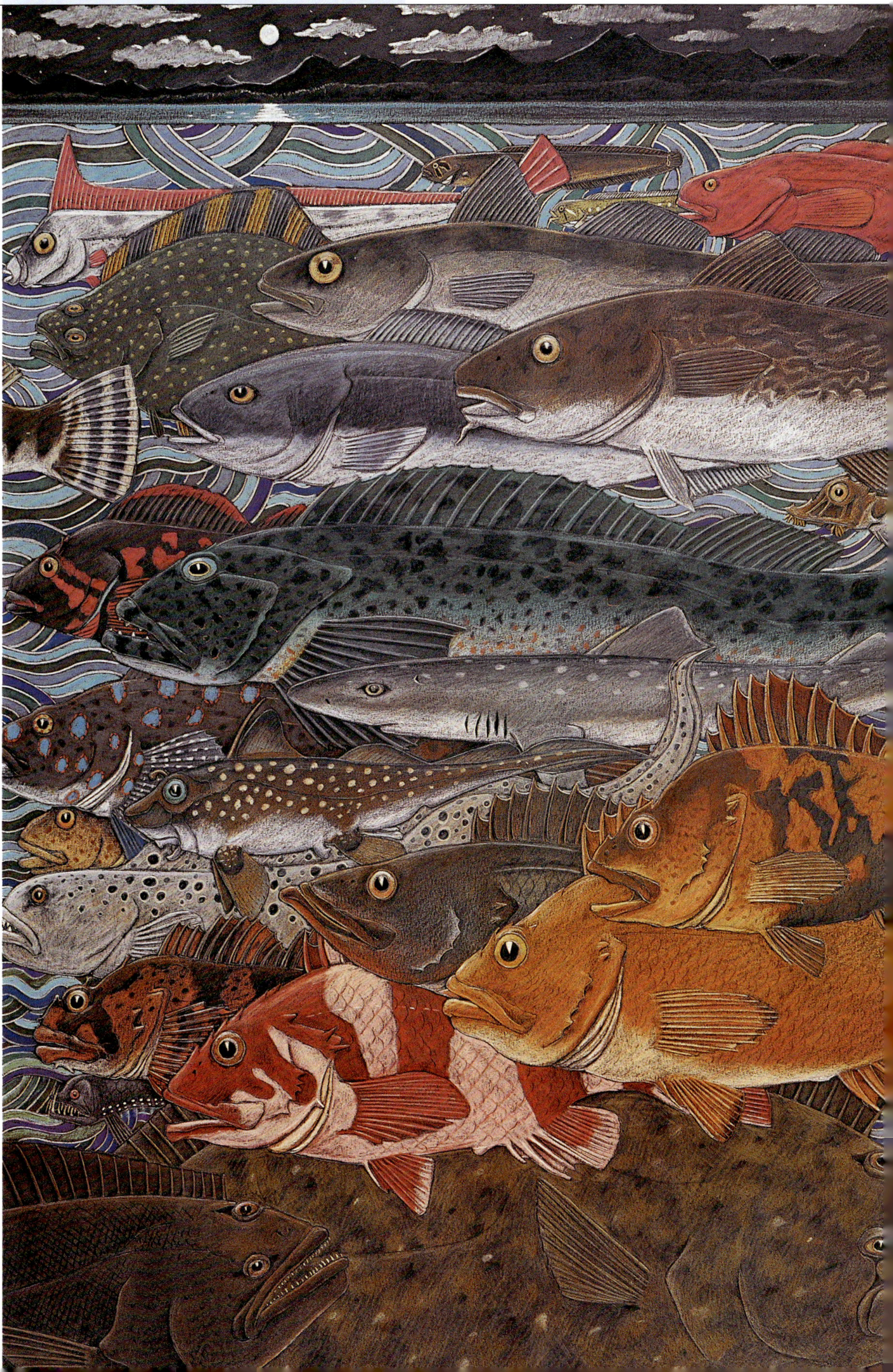

THIS IS one of my earliest colored-pencil drawings of "all-Alaskan" fish. I was a piscatorial novice at the time and was learning about the different terms applied to fish along the Pacific coast. Bottom fish prefer deeper water and tend to hang out within a few feet of the ocean floor, either on sandy substrates or around rocky pinnacles. I loved the sheer biological overload as I dove deep into the ID books and hung around the docks watching boats being unloaded. My studio was over a fish processing plant at the time, and I could literally run downstairs to see what the latest catch was. My sister Kate and brother Tim helped launch me on my publishing career in 1984 when they put up the bucks to print this as a poster. I paid them back in a year's time. The posters are long out of print, but I sometimes run into a faded copy of one in a fellow fish head's office or home.

1987, charcoal on paper, 40 in. × 26 in.

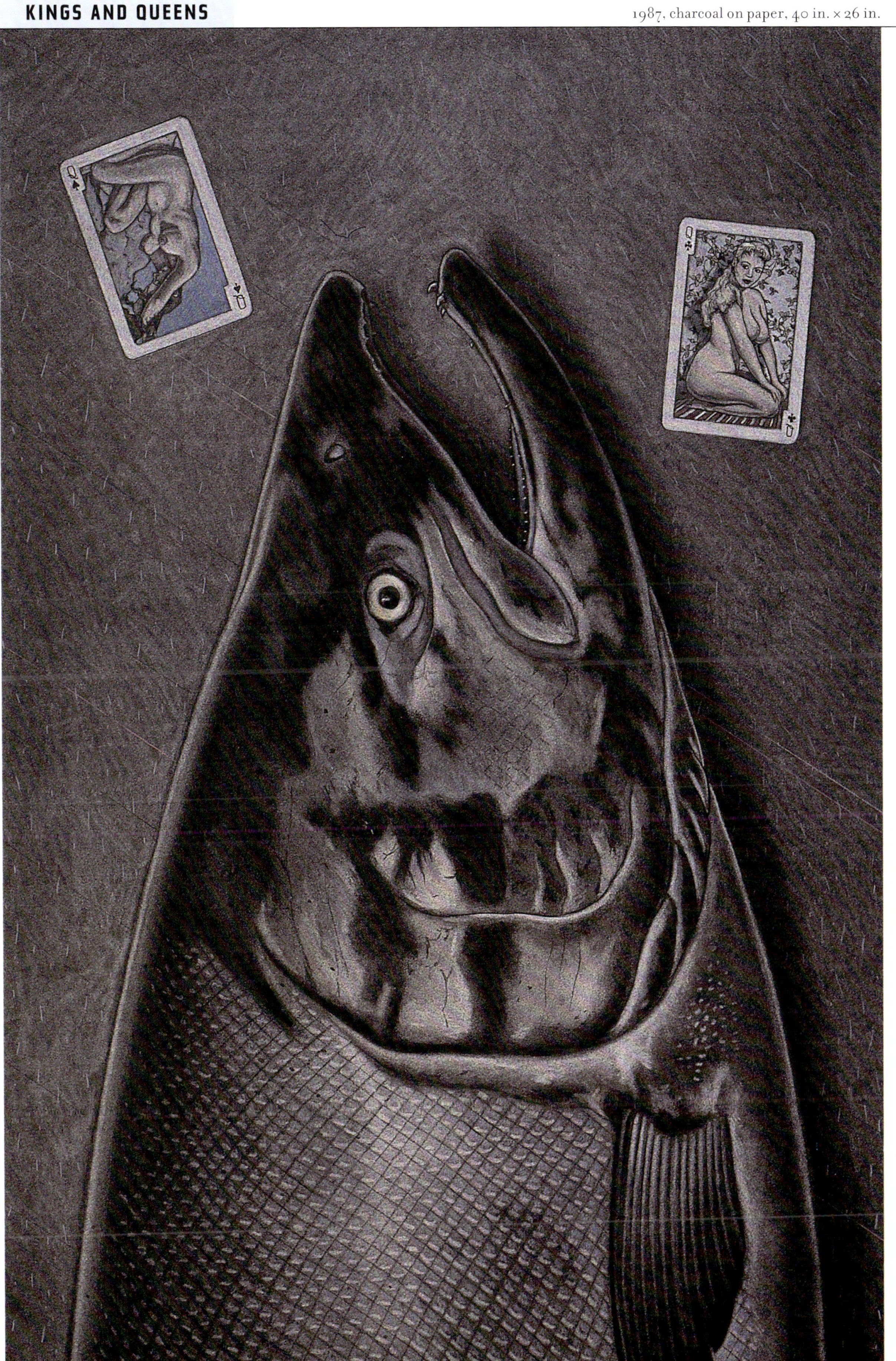

ANOTHER PUN with sexual overtones visualized.

FISH WARS

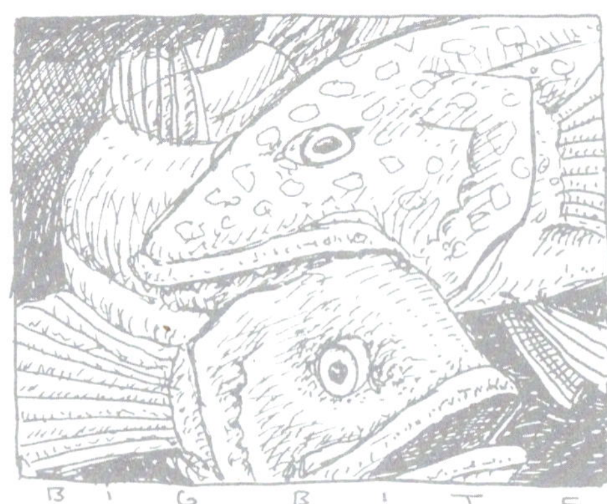

I USED a very different technique for this drawing, transferring Xeroxed images with a powerful brain-numbing solvent called xylol onto a piece of archival drawing paper, much as Robert Rauschenberg did with magazine images in the 1960s. I scavenged the pictures from old technical books, then painted in the background with enamel spray paint and colored the fish, German WWII airplanes and buzz bombs, and the guy in the gas mask with colored pencils. The process involves lots of burnishing with a spoon on the backsides of the copies.

Probably because I grew up on air force bases, planes are the second love of my sketchbook life after dinosaurs, and I spent much of my youth drawing endless versions of aerial dogfights. This drawing harks back to my childhood impulses but now includes fish participating in a battle in a galaxy far, far away.

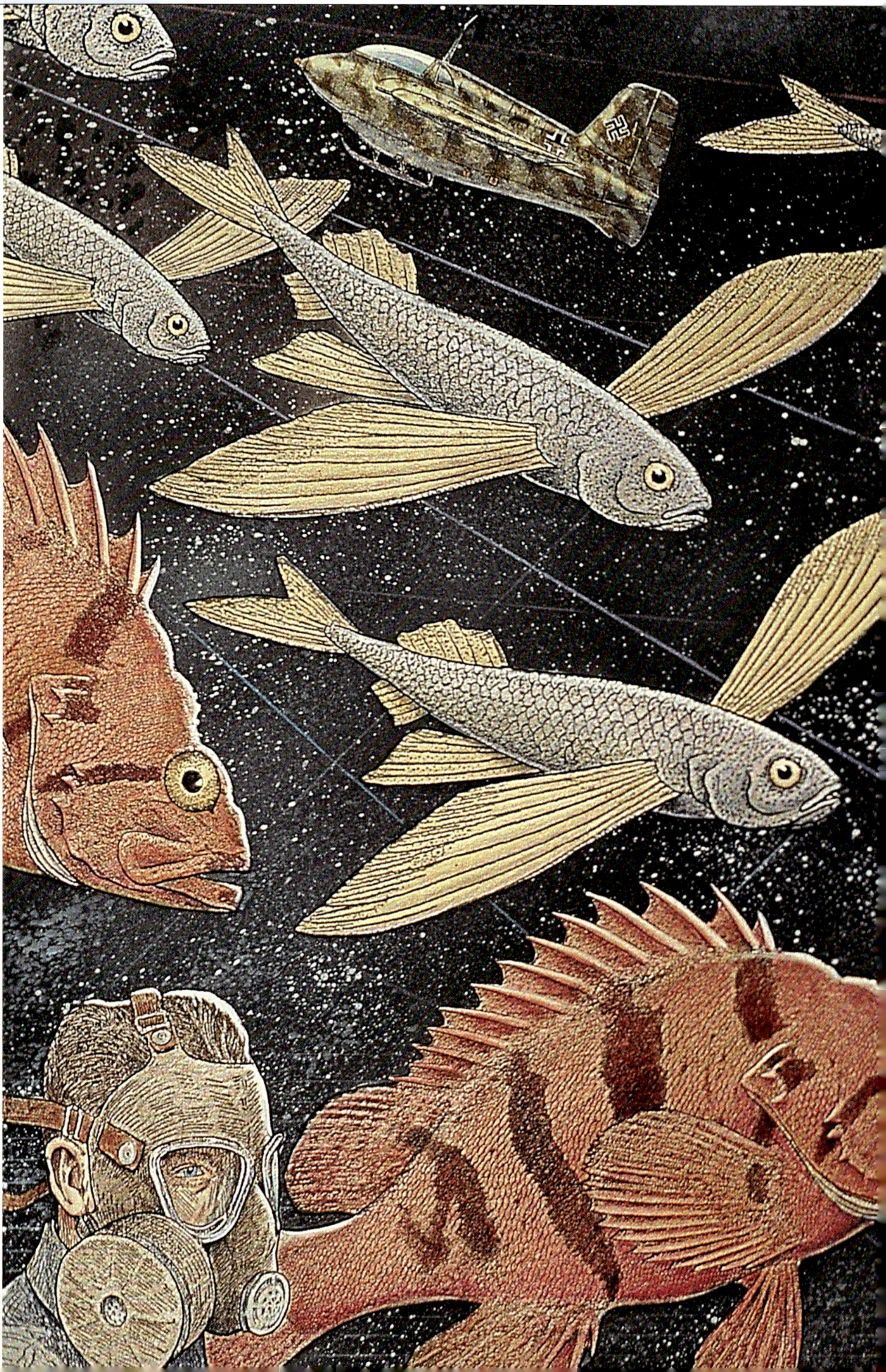

1985, mixed media on paper, 30 in. × 38 in.

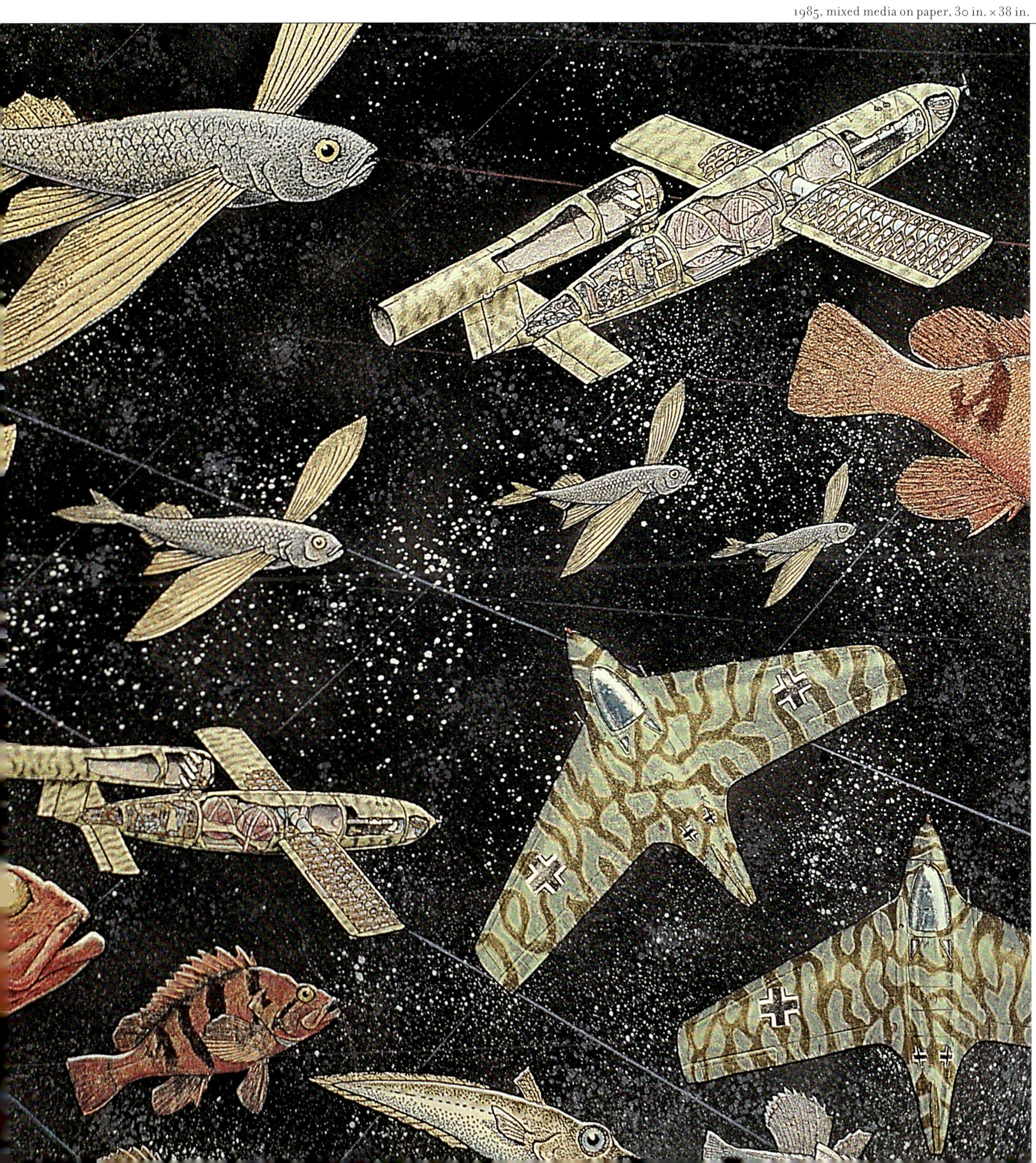

ROCKFISH

1998, acrylic on canvas, 48 in. × 36 in.

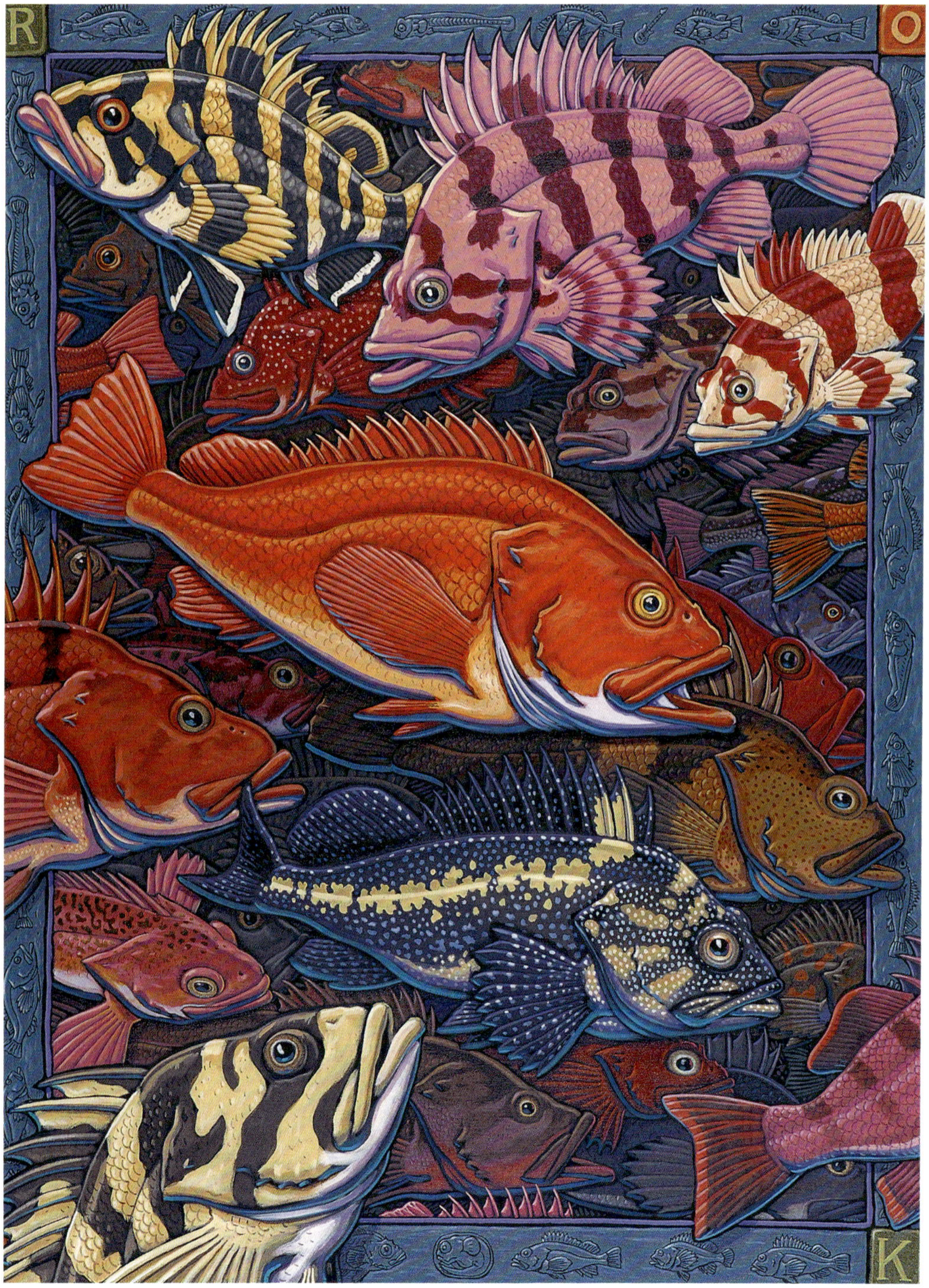

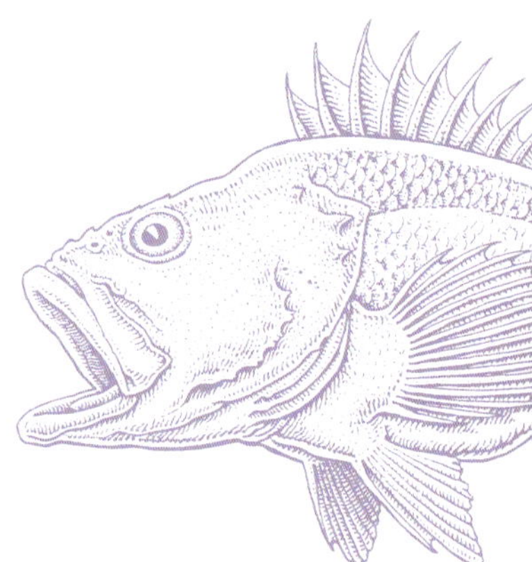

Dr. Milton Love, a friend from UC Santa Barbara and *the* stand-up comedian of the ichthyological world, approached me in the early 1990s to paint a cover for his definitive book on the rockfishes of the Northeast Pacific. Milton had been working on this opus for nearly a decade with two coauthors. I was honored and excited to work with Milton because his tongue-in-cheek humor appeals to my warped sensibilities. He ended up putting together a science book full of art, humor, and facts about these "magnificent" (the meaning of their genus name, *Sebastes*) fish.

Milton gave me carte blanche in designing the cover, provided of course that the fish depicted were all *Sebastes* species. I filled a canvas with beautiful rockfish, and painted larval and juvenile forms in the border area along with an electric guitar (they are ROCKfish after all). The letters in the corners are meant to spell out R-O-C-K, but the C is obscured.

1995, pastel on paper, 32 in. × 32 in.

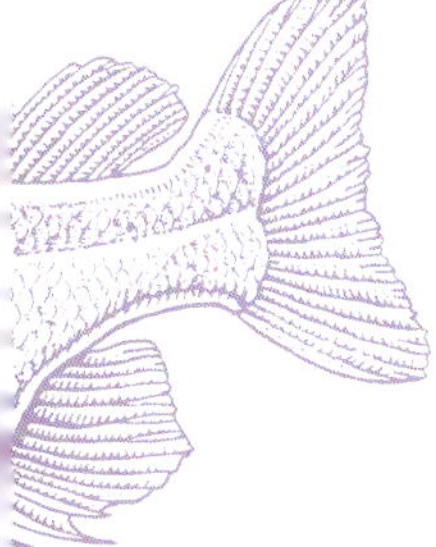

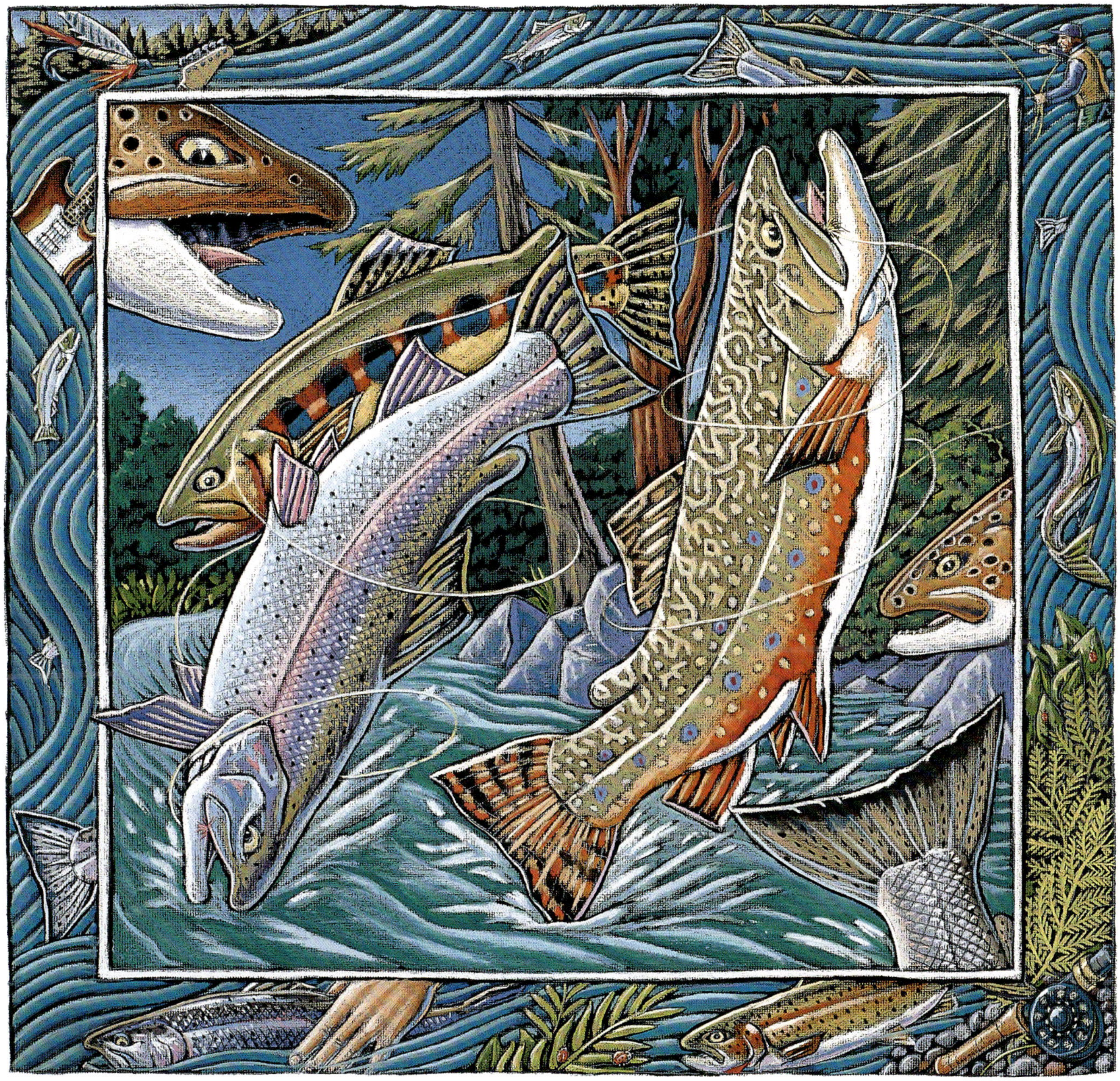

The wily trout is sacred quarry for fly fishermen—though fly fishing is an avocation I have honestly yet to master. My amazing ability to immediately tangle a fishing line is probably at the heart of this drawing. You can spot four ladybug beetles in the border as a subtle reference to the bad pun, based on an old Beatles song, that inspired this image. There's a Stratocaster guitar in there as well. The pantheon of "sacred" trout represented are brook, rainbow, apache, brown, and the tail of a steelhead.

WEAPONS OF BASS DESTRUCTION

1995, pastel on paper, 30 in. × 28 in.

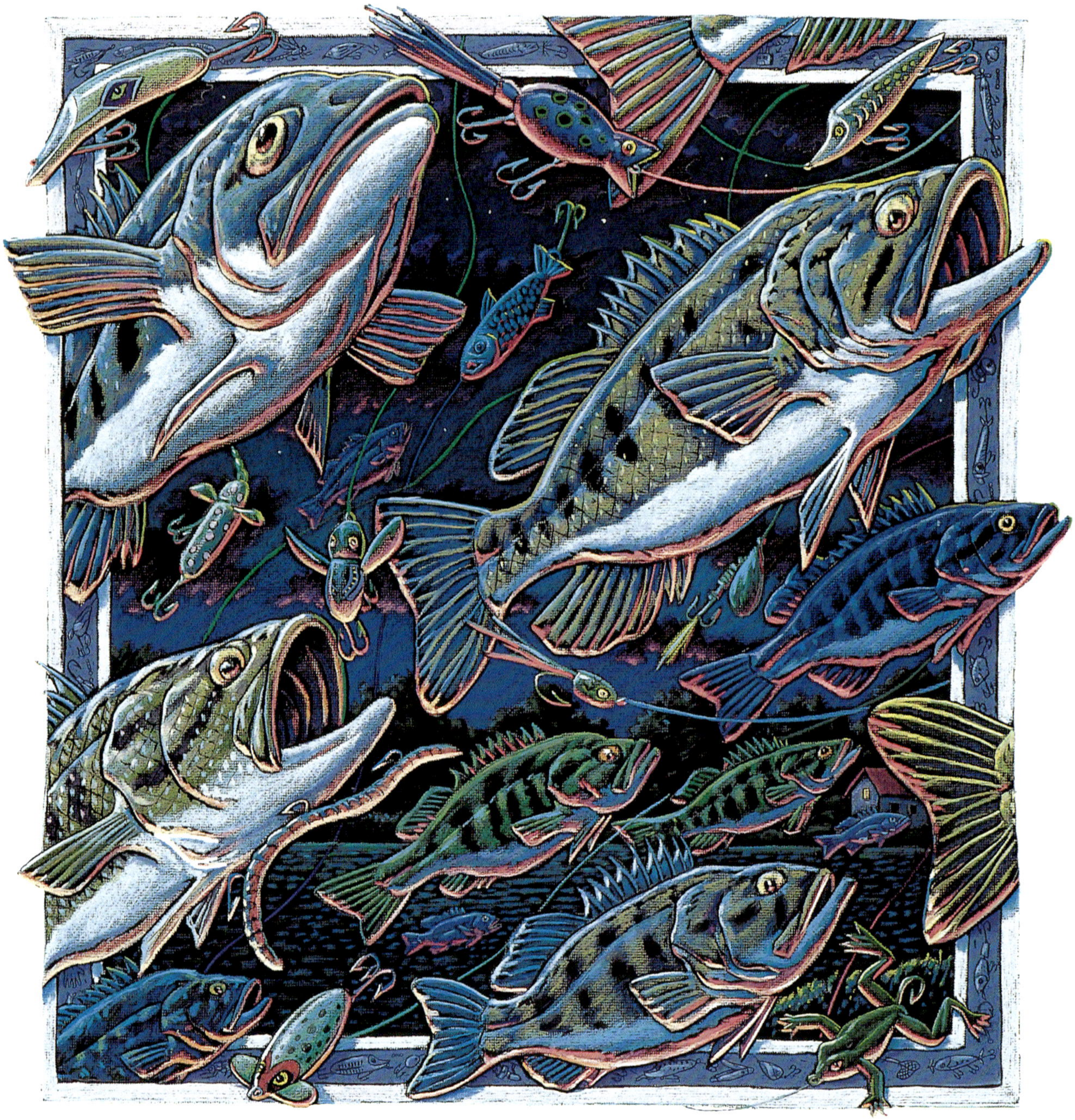

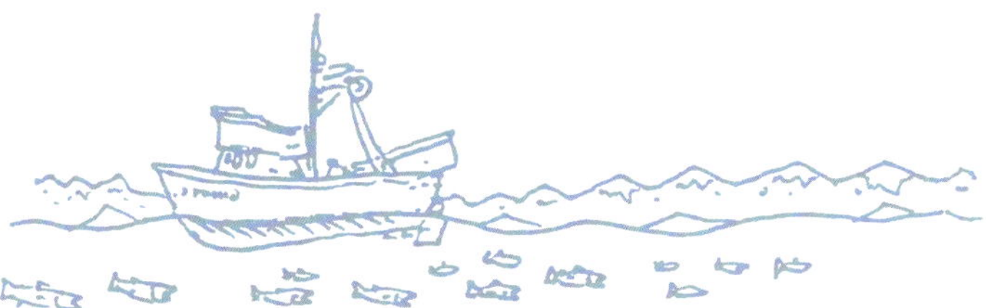

THE TITLE IS a bad pun by any stretch of the imagination, but these days it's an unusually timely one. Vintage fishing lures have fascinated me for a long time, and I had a great time sprinkling them throughout this drawing.

DREAM OF THE FISHERMAN

1987, charcoal on paper, 22 in. × 30 in.

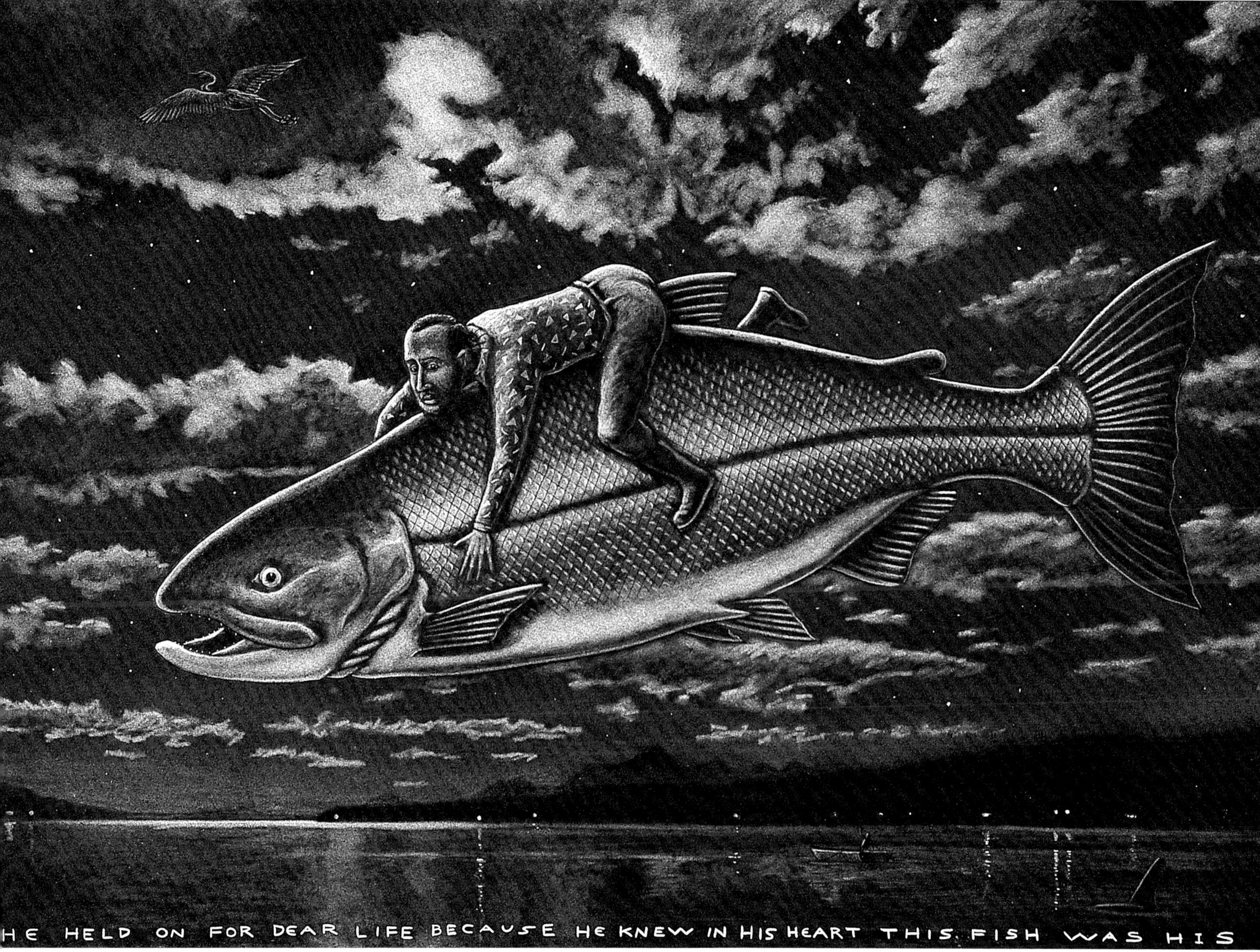

I DID THIS drawing shortly after I caught my very first king salmon, a forty-one-pounder that fought me for over an hour. When I finally got it into the boat I truly understood the pure adrenaline-fueled joy of hooking into a monster fish: a lunker of that size makes you never want to let go. The determined fisherman attached to the back of the fish was inspired by a sculpture I'd seen of an Inuit shaman riding on top of a walrus. I took a roll of photographs of a moonlit sky over the Tongass Narrows specifically for this drawing.

1989, pen, ink, and watercolor on paper, 8 in. × 8 in.

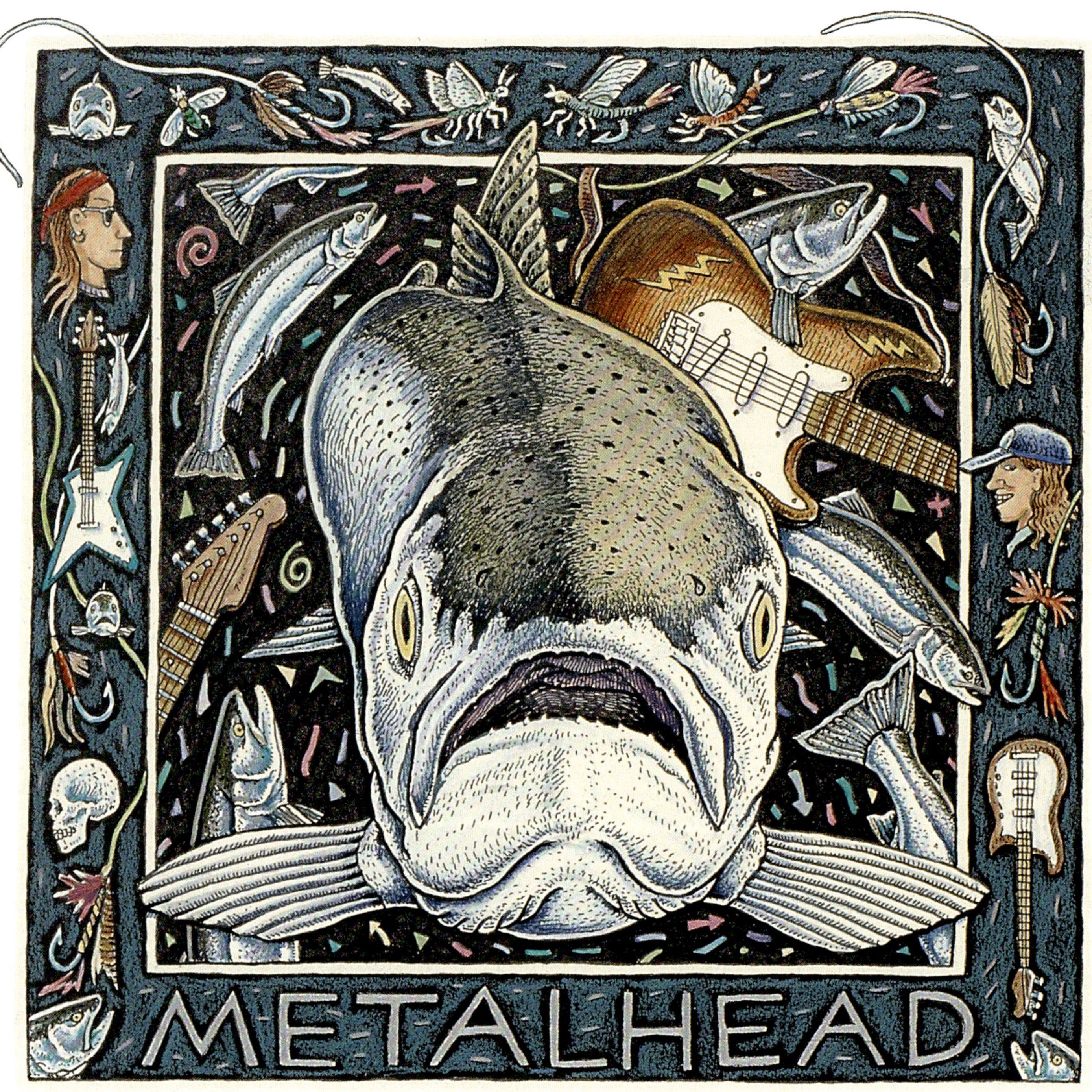

Steelhead trout are called "metalheads" in some circles. For me, the name conjures visions of heavy metal head-banging music.

REBEL WITHOUT A COD

1990, pen, ink, and watercolor on paper, 11 in. × 9 in.

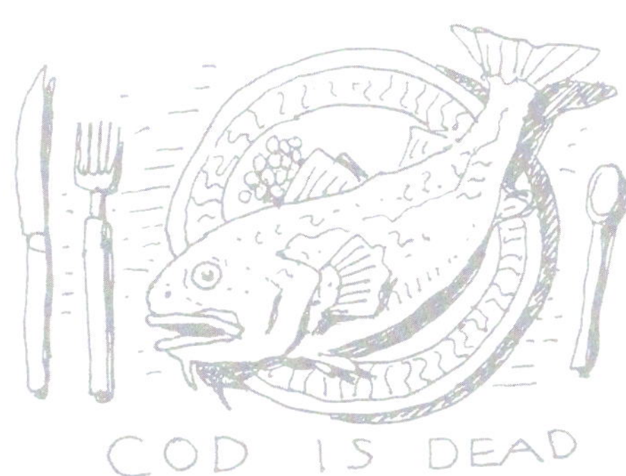

THIS PUN cried out for an image of the original rebel James Dean. We printed up a run of a dozen prototype shirts, then sent the representative of the James Dean estate a copy of the design and asked permission to use his image or to arrange for a licensing deal. After a few weeks we got a polite "no thank you" note back and we killed the project. I'm not sure what ever happened to those shirts. I later did a version of the design without James in it. That one featured a friend of mine named José Mateu seated on his Harley in full leather biker gear sporting a fishing rod.

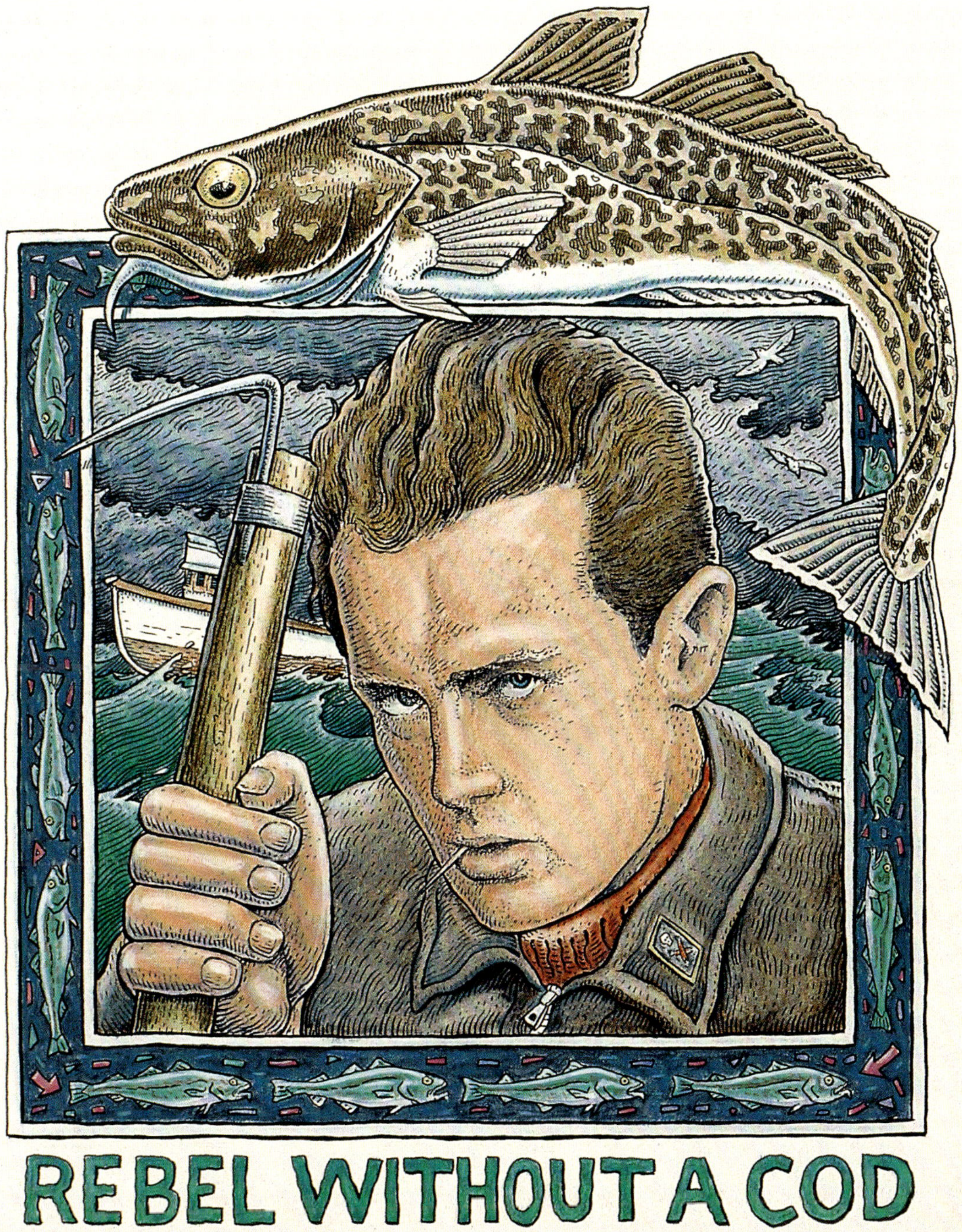

REBEL WITHOUT A COD

1988, pen, ink, and watercolor on paper, 8 in. × 6.5 in.

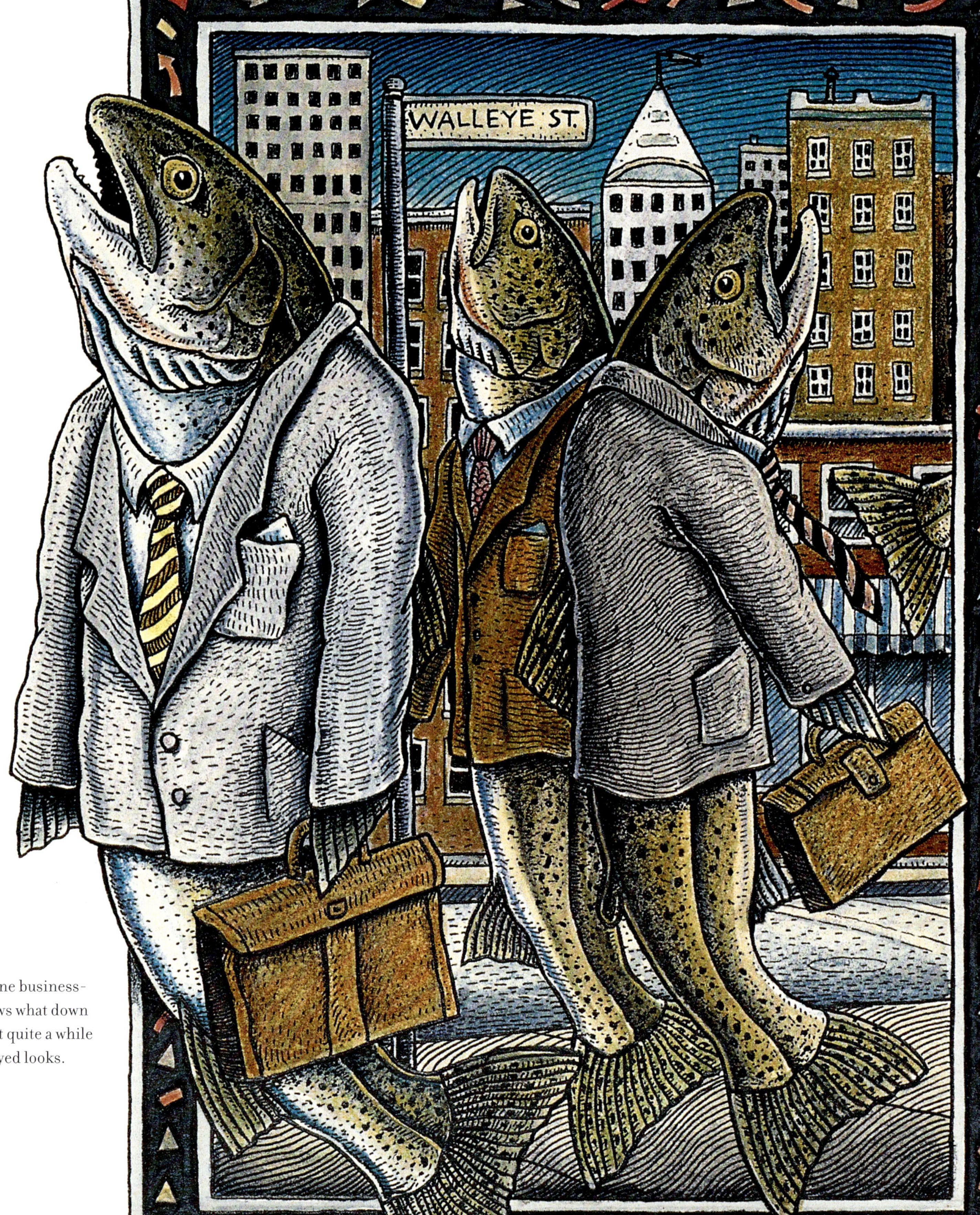

THREE NEFARIOUS piscine businessmen doing god only knows what down on Walleye Street. I spent quite a while perfecting their shifty-eyed looks.

BASSACKWARDS

1988, pen, ink, and watercolor on paper, 12.5 in. × 11 in.

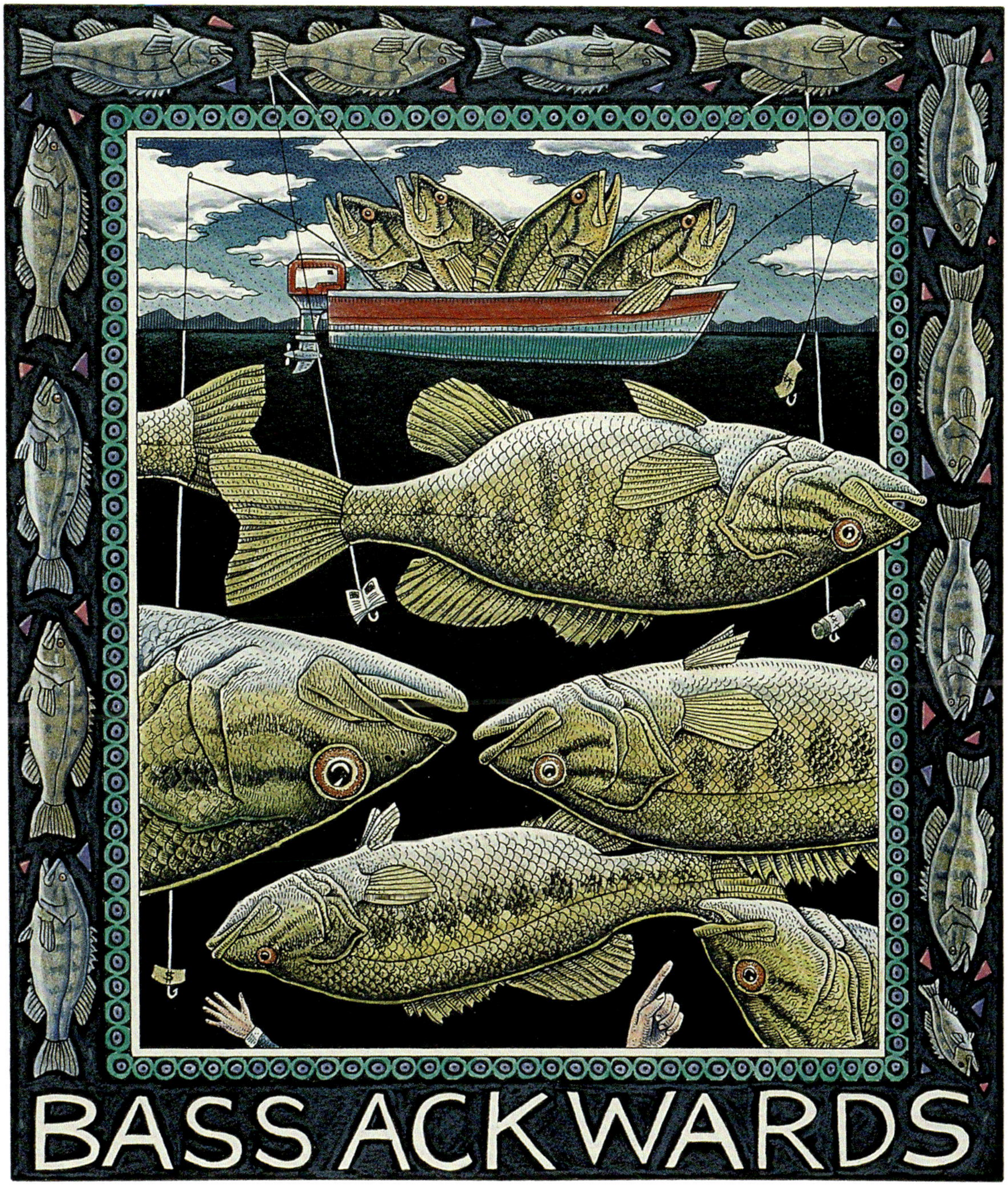

THIS HAS PROVEN to be one of my more popular shirts over the years. I liked the idea of the fish going after the humans, using things like money, liquor, and—much less successfully—literature as bait.

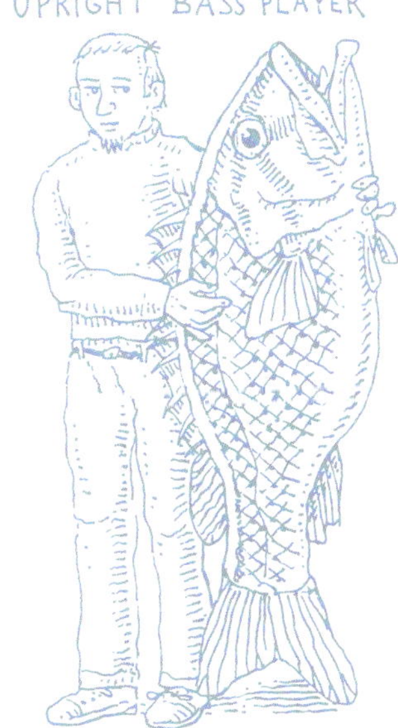

1991, pen, ink, and Photoshop color, 10 in. × 8 in.

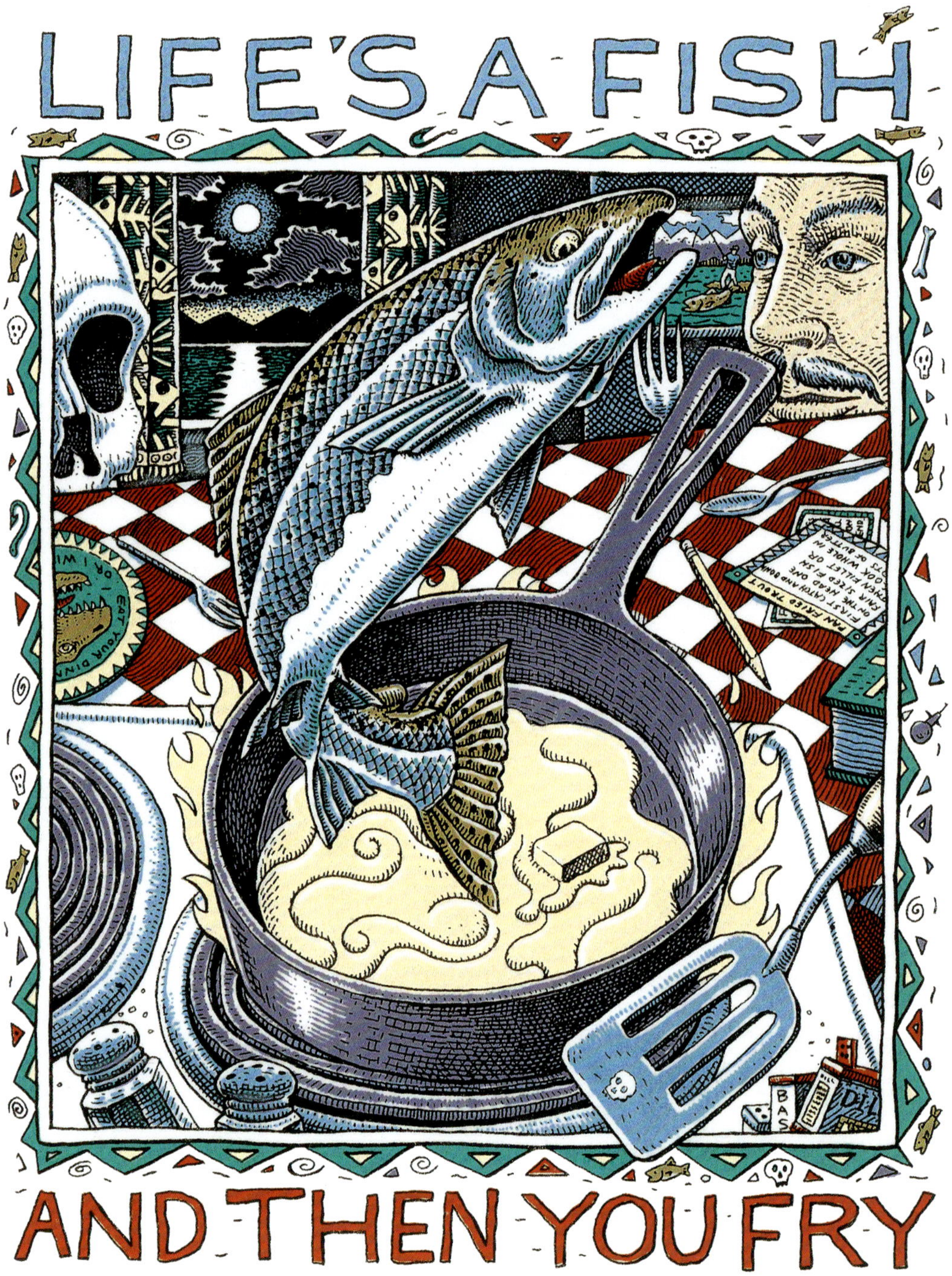

LIFE'S A FISH

AND THEN YOU FRY

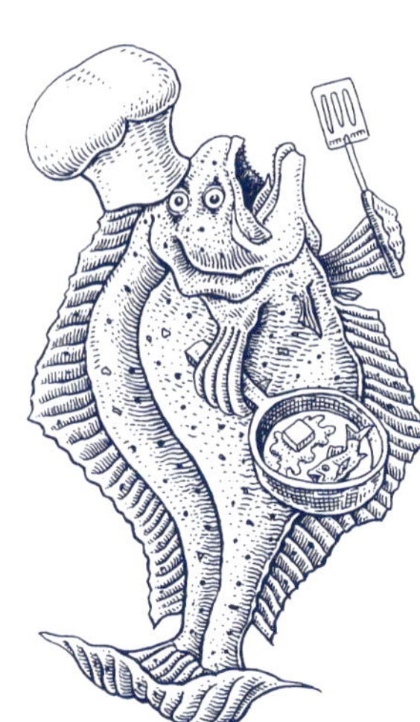

THIS PUN was suggested to me by my kid sister Susie. It later became the cover for an Alaskan seafood cookbook written by Randy Bayliss of Juneau. This is definitely not a "catch and release" kind of shirt. Check out my secret to tasty fish on the recipe card on the table.

1988, pen, ink, and watercolor on paper, 12 in. × 12 in.

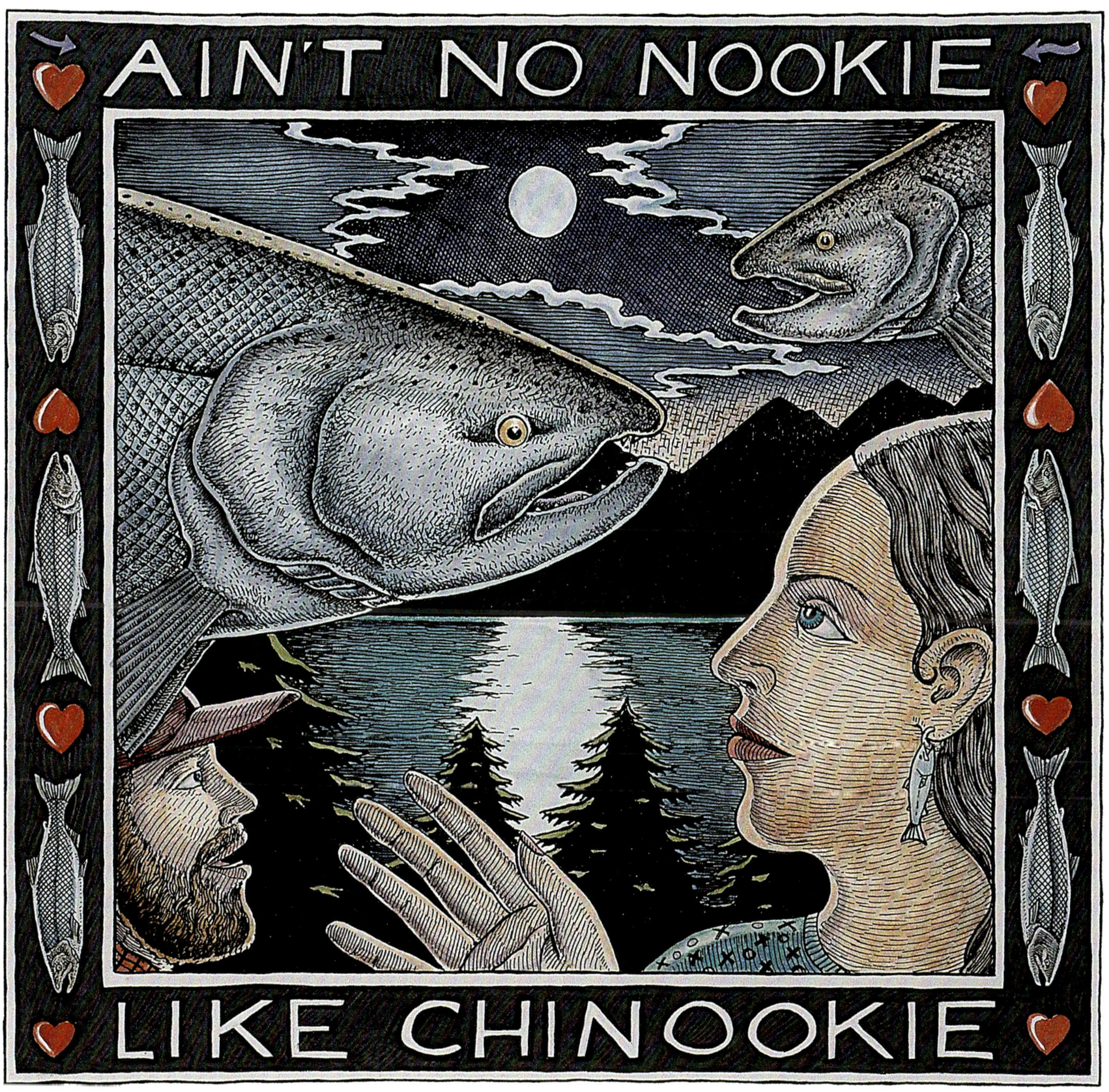

I DON'T THINK IT'S REALLY quite true that catching a fish compares to the ecstasies of love, but it's fun blurring the line between the two. Bob Waldrop, an impresario of the Alaskan fish processing world and an avid collector of humpy art, tossed this phrase out to me one beer-soaked evening in a Ketchikan bar.

HELL'S ANGLERS

1988, pen, ink, and watercolor on paper, 12 in. × 10 in.

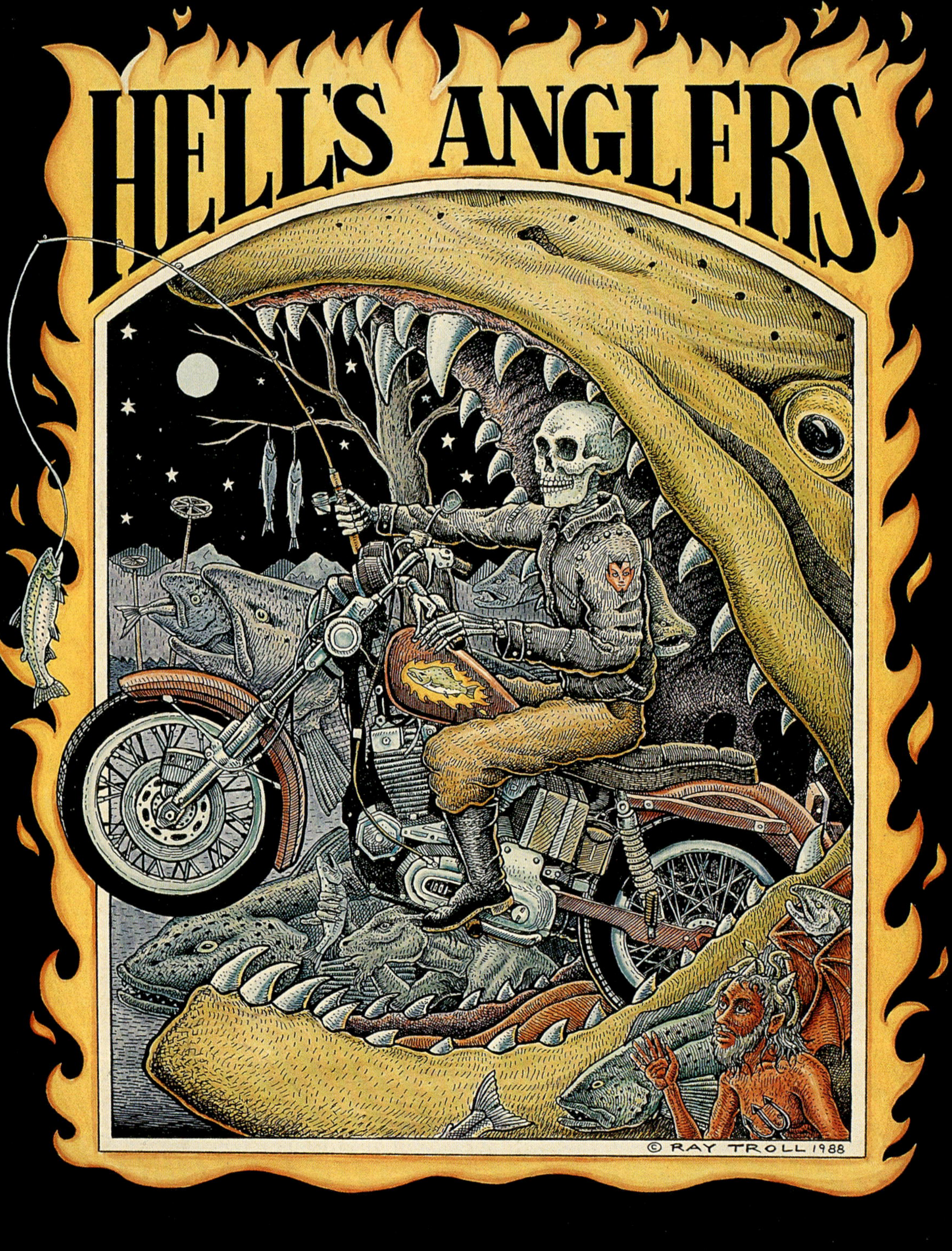

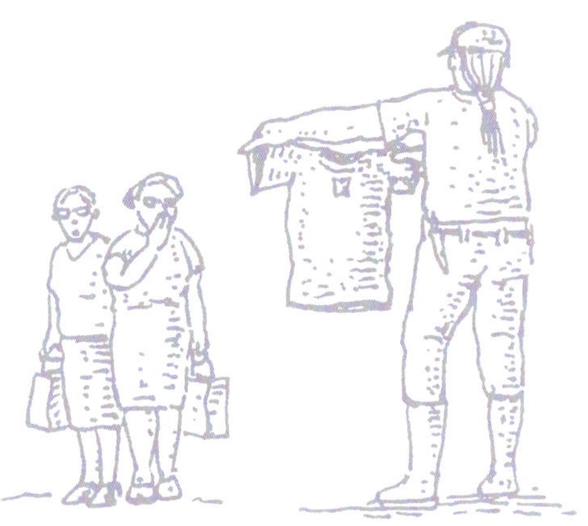

In the spring of '88 I received a letter from a gang of fly-fishing bikers from the Homer, Alaska, area who called themselves the Hell's Anglers. They asked me to do a shirt for them, and I gladly said I would because it was such a cool name, but only if I could add it to my commercial T-shirt line. I dragged out my art history books and made a drawing that refers to Bosch's and Bruegel's depictions of hell.

A few months after the shirts came out I got a call; apparently, some retailers in northern California had received negative feedback from a few of the local Hells Angels groups. I didn't want to get into a disagreement with them, so we immediately stopped production. The shirts quickly became a much-coveted collector's item.

1987, serigraph, 30 in. × 22 in.

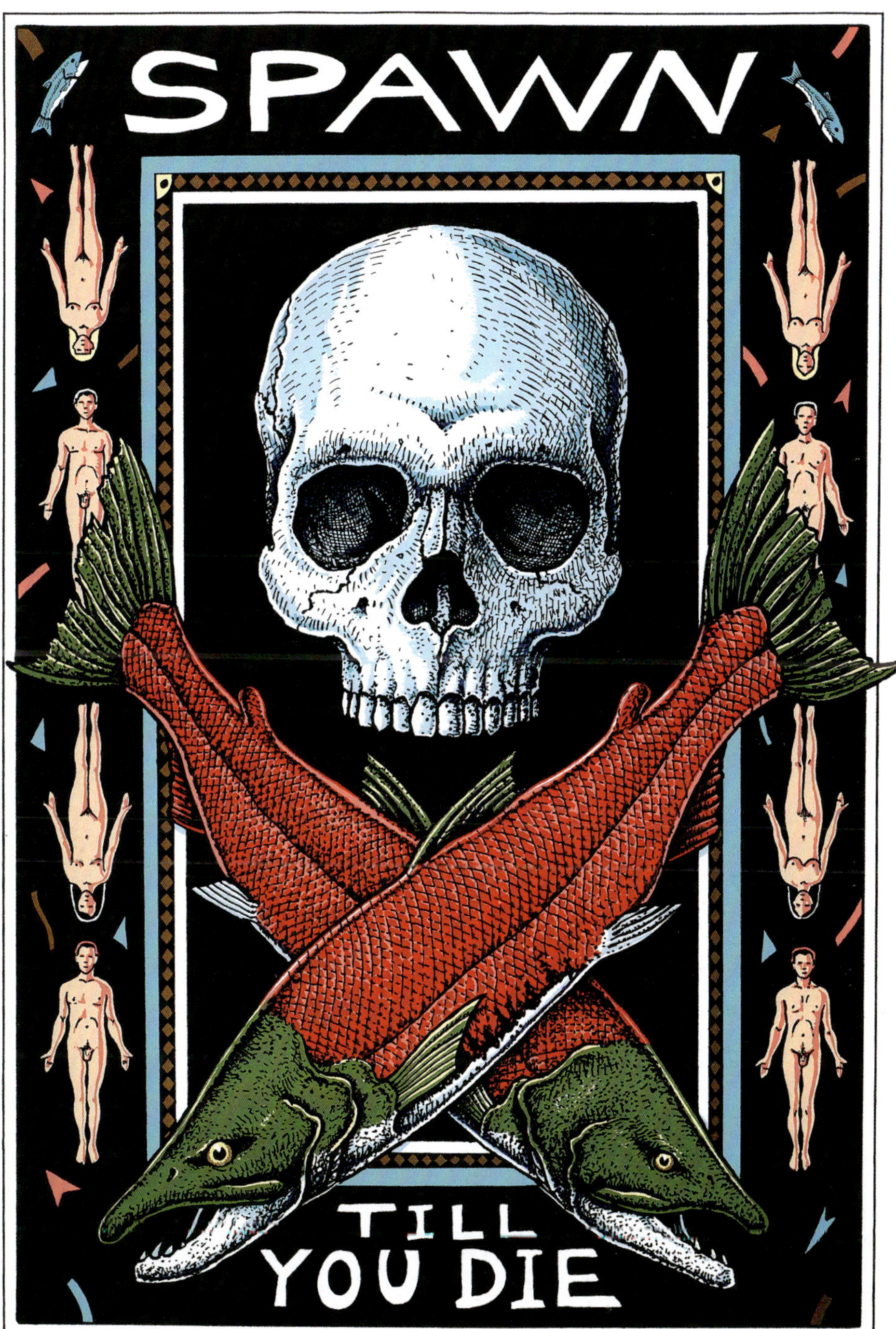

This seems to be the image that most people remember me for. It's been the most popular T-shirt and poster for over a decade and would probably make a nice epitaph for my tombstone. It sums up the two great urges in life quite nicely: Eros and Thanatos; Love and Death. What else is there? They connect us to all our vertebrate cousins and by extension to all the spineless wonders as well.

The shirt shows up in movies every now and then, inevitably worn by a bad guy—and it's usually one of the classic black-and-white shirts we first produced. I've spotted it on disaffected teenagers in *Newsweek* magazine and on the occasional rock star in *Rolling Stone.* One of my proudest moments was seeing a member of Metallica wearing one in a *Hit Parader* magazine spread.

An earlier version featured dead salmon and human skulls on a creek bank, but it didn't quite congeal as a design until Bill Spear, the king of enamel pins in Juneau, casually suggested more of a Jolly Roger approach. That captured the zeitgeist of the Northwest just right.

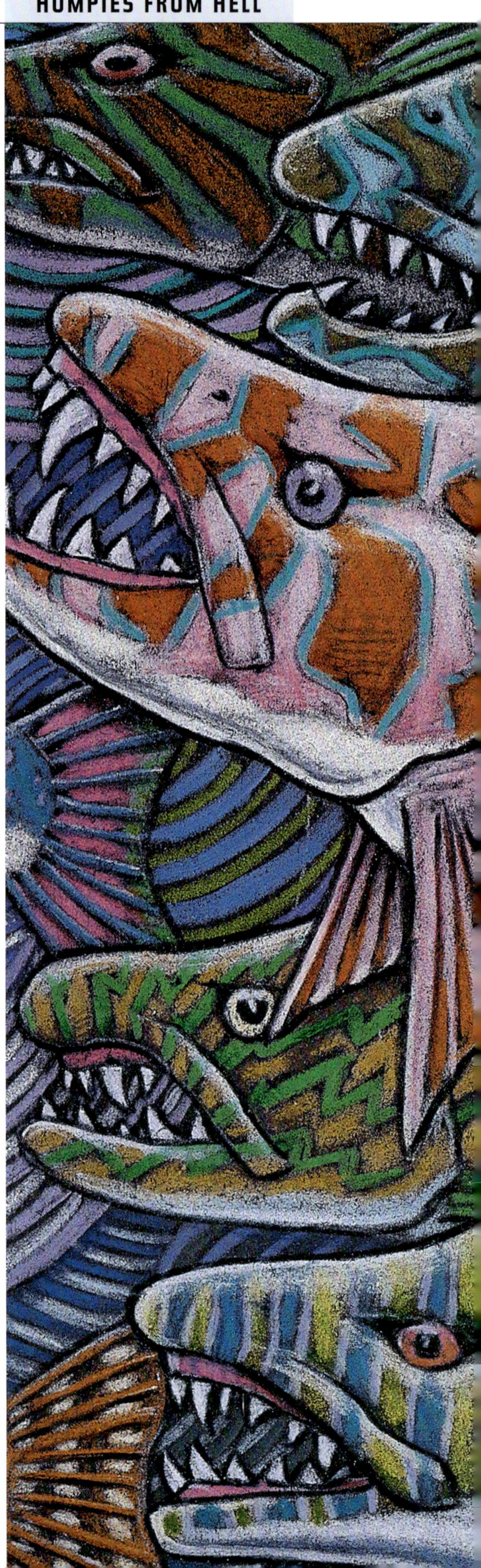

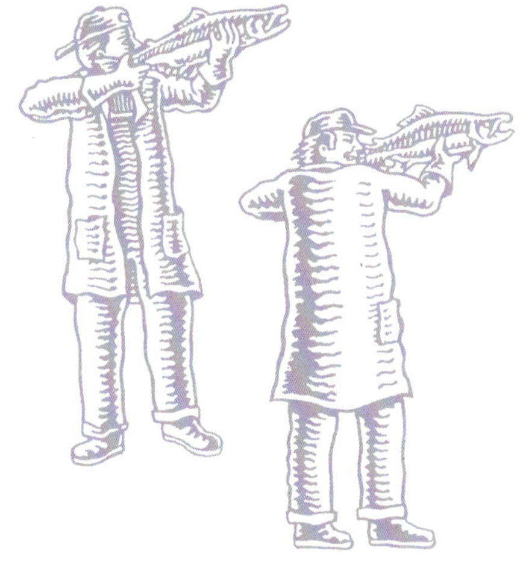

Salmon transform themselves from sleek, silvery oceangoing fish to sexually obsessed humpbacked, snaggle-toothed monsters when they enter freshwater streams to spawn. It's a terminal trip, and they have only one shot at mating and their link to eternity.

Humpies (a.k.a. pink salmon) are generally viewed as a mixed blessing on the north coast. They arrive in late summer in such overwhelming numbers that they seem like a biblical plague upon the land as they fill every available ditch, gully, stream, and river. By the fall their rotting bodies litter the banks of local streams and reek to high heaven. Commercial fishermen weary of them as they catch countless tons of them. Being the smallest of the five commercially harvested species, humpies are always ranked as the least desirable and are worth the fewest dollars per pound. Sport fishermen also tend to hold them in low esteem. So it's not too much of a stretch to add the phrase "from hell" to their name.

1984, colored pencil on paper, 7.5 in. × 9 in.

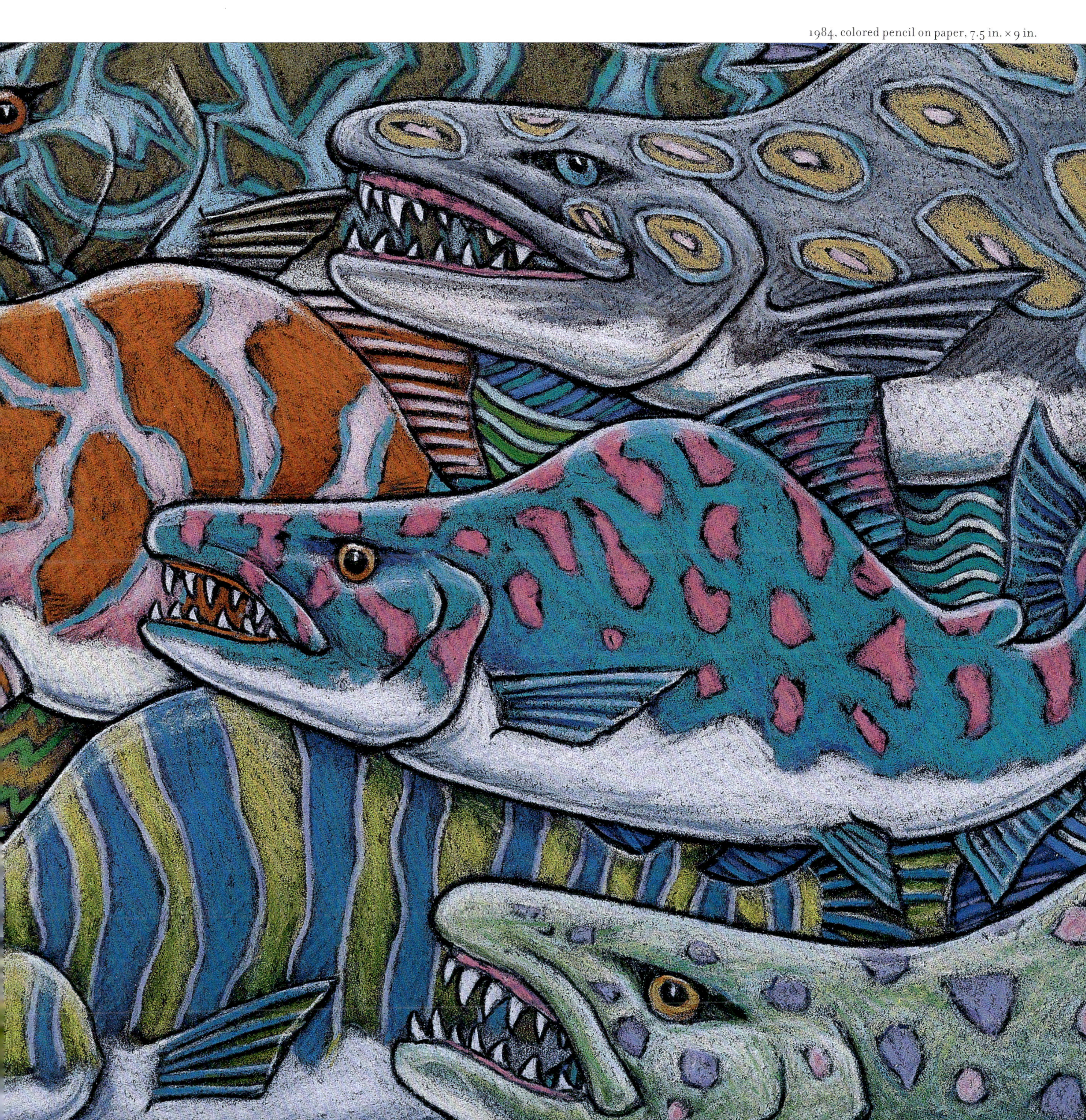

FISH HEAD

1990, pastel on paper, 22 in. × 17 in.

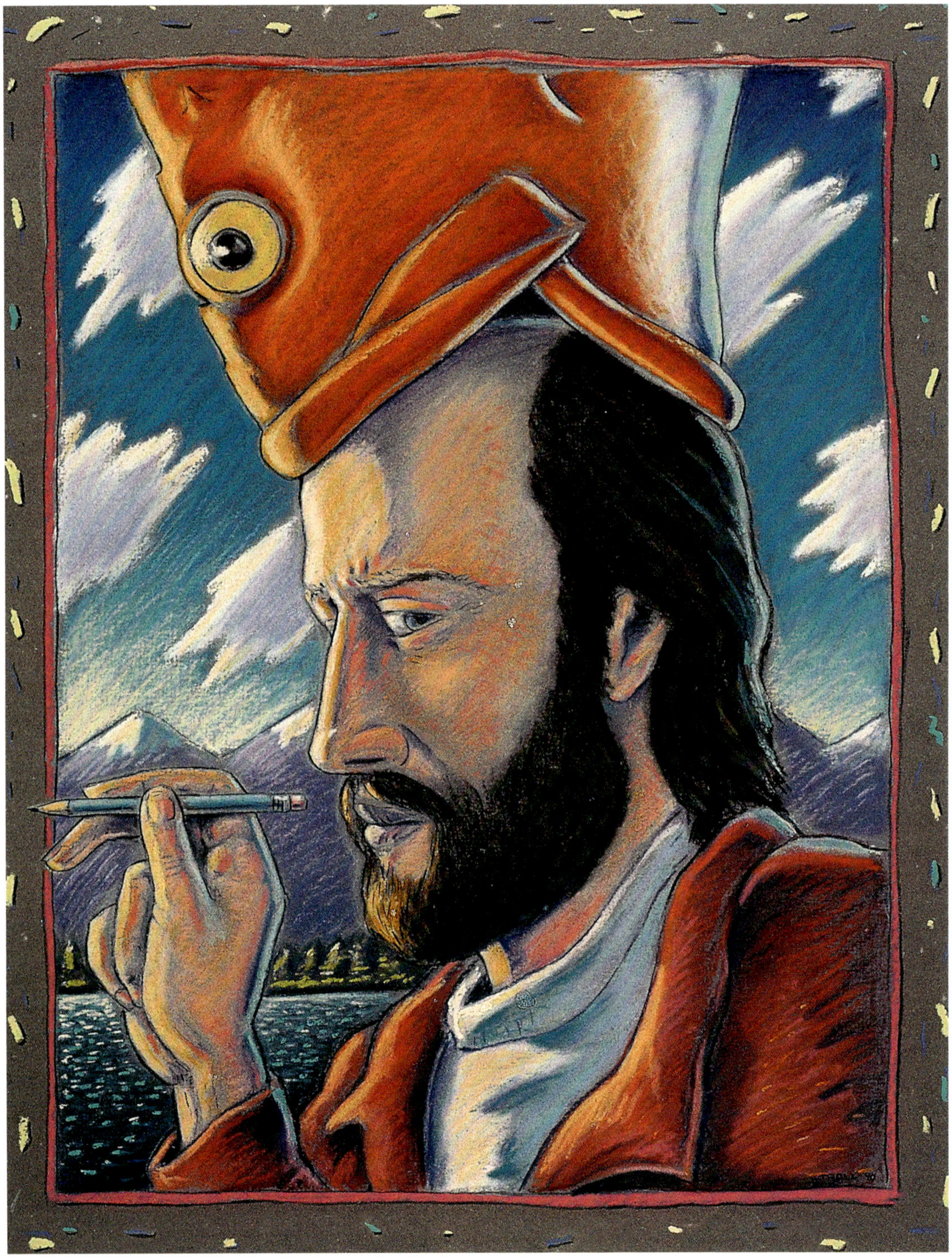

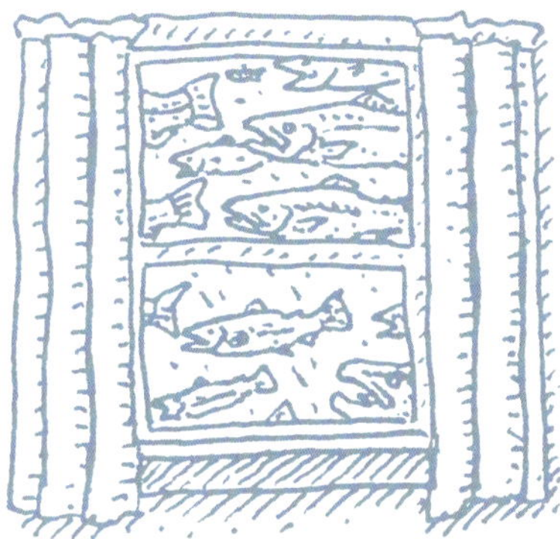

THE FISH ATTACHED to my head in this self-portrait is a large red snapper. I've come to appreciate several local Alaskan fish more than others, and this particular rockfish is high on my list. The scientific name is *Sebastes ruberrimus*, which means "magnificent, very red," although they really are a bright orange. With their neon color and big yellow eyes, they are the very definition of a cartoon fish and show up often in my work. Because reds shift to black as the color spectrum fades in the depths, their orange color camouflages them.

When fish iconography began to take over my artistic life I decided to roll with it, and this image sums it all up: it's the ultimate fish-mind meld. I was inspired by small Tlingit wood carvings from the nineteenth century that depict human beings popping out of the mouths of yelloweye rockfish. When rockfish, which regulate their buoyancy with sensitive air bladders, are brought quickly from deep water to the surface on the end of a fishing line, their air bladders fill with gas, protruding grotesquely through their mouths once they're landed. I think the Tlingits were portraying this phenomenon, and I followed their lead.

SOMETIMES LATE AT NIGHT . . .

1984, pen, ink, and colored pencil on paper, 8 in. × 12 in.

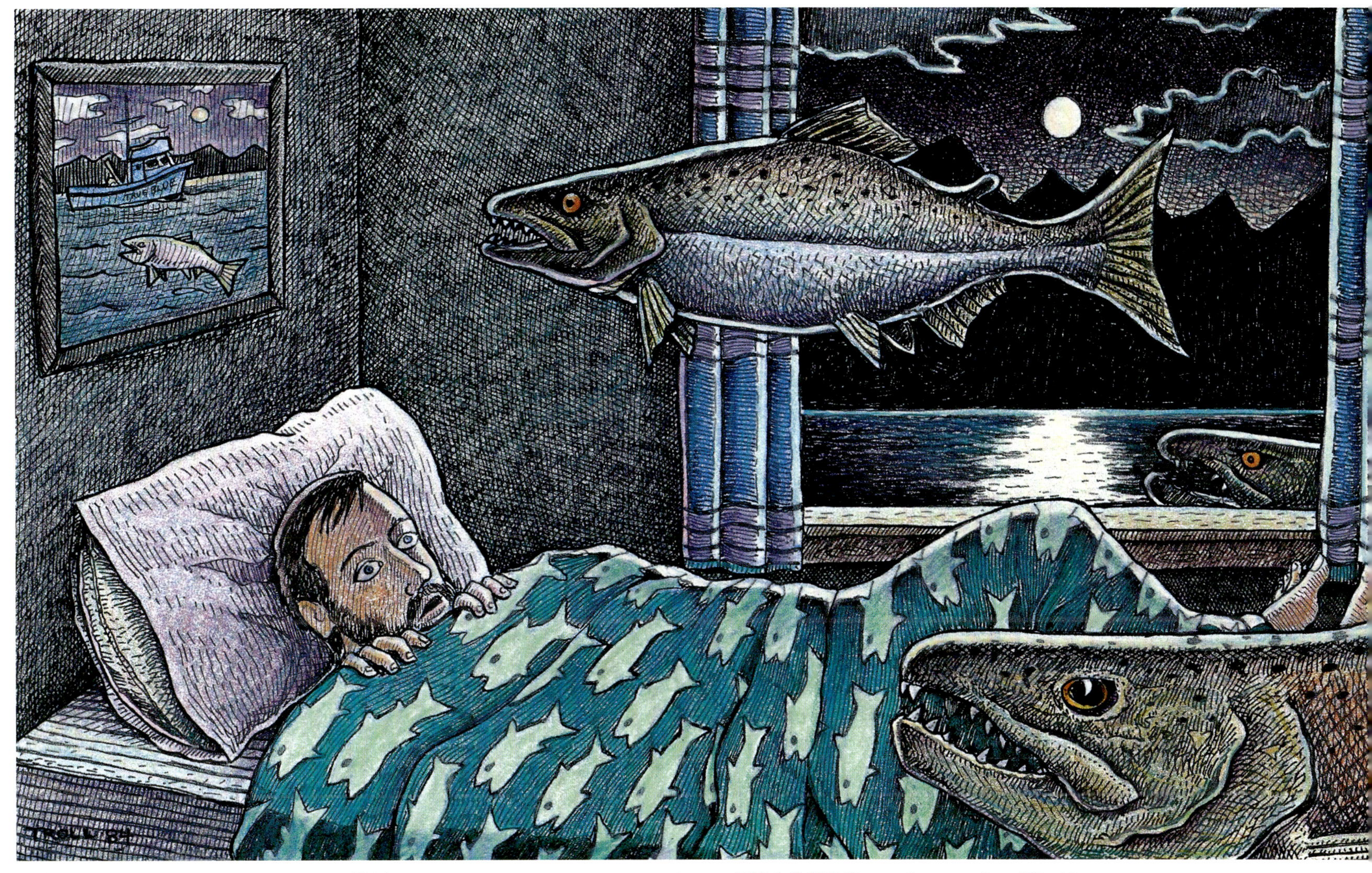

SOMETIMES LATE AT NIGHT THE SPIRITS OF ALL THE FISH I'VE EVER CAUGHT COME BACK TO HAUNT ME.

I WINCE each and every time I clobber a fish on the head; it truly weighs on my soul. "But as long as I eat what I catch, and the fish population isn't threatened in any way, and the fish is dispatched quickly"—that's my running inner mantra as I try to assuage my guilt. I've had scores of fishermen tell me they suffer the same pangs of remorse. It's a dilemma best solved with garlic, lemon, and butter.

1989, pen, ink, and watercolor on paper, 7.5 in. × 9 in.

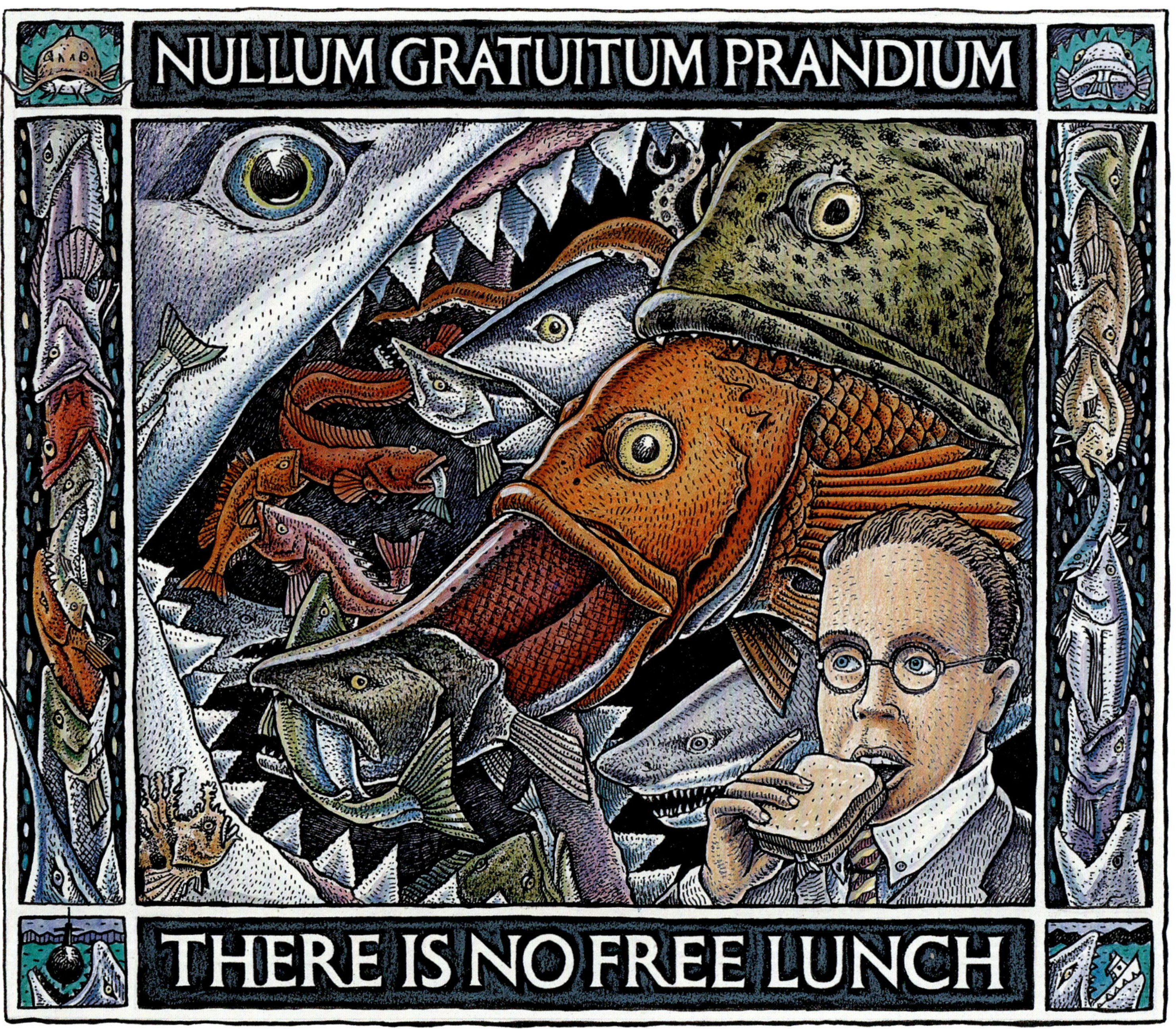

THIS IS my homage to Pieter Bruegel's 1556 masterpiece *Big Fish Eat Little Fish*, one of his better-known depictions of Dutch proverbs. I gave it a little bit of a Southeast Alaskan twist with ling cod, snapper, decorated warbonnets, and sockeye salmon. Humans are a big, big part of the food chain these days, so I included a nerdy-looking fellow snarfing a sardine sandwich. I've always liked the border on this one and was delighted to see it tattooed around the wrist of an avid Troll fan.

1990, pen, ink, and acrylic paint on scratchboard, 7 in. × 8.5 in.

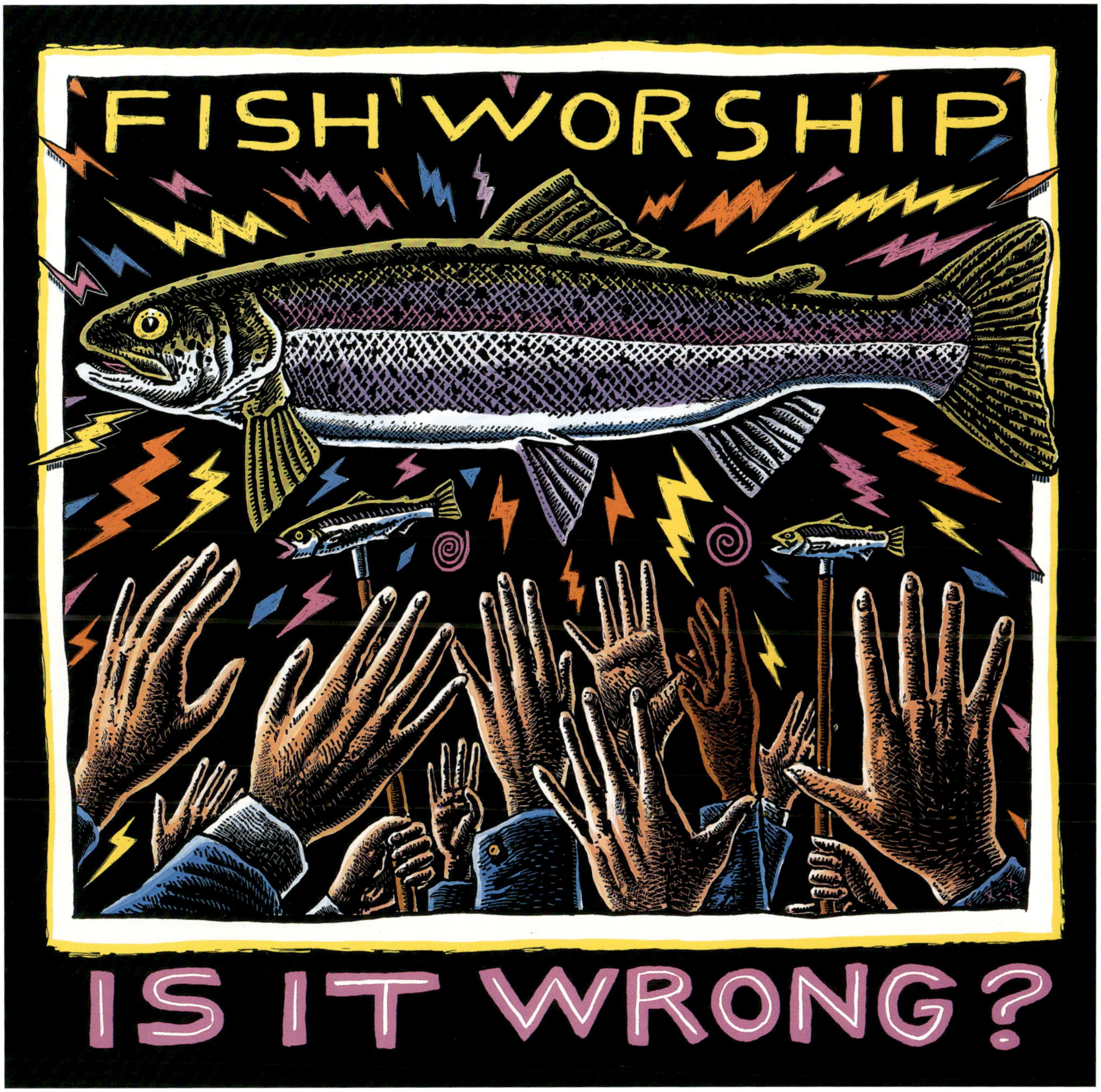

THIS ONE popped into my head one day fully formed. It doesn't really make much sense, but that's the best sort of image sometimes. If you're a fish-obsessed fanatic, you understand this one on a visceral level.

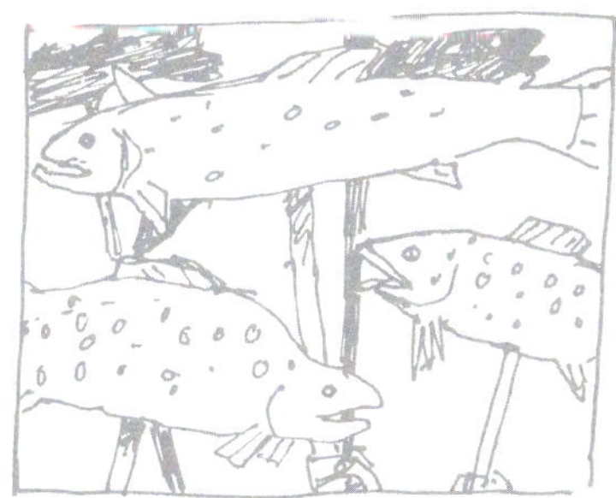

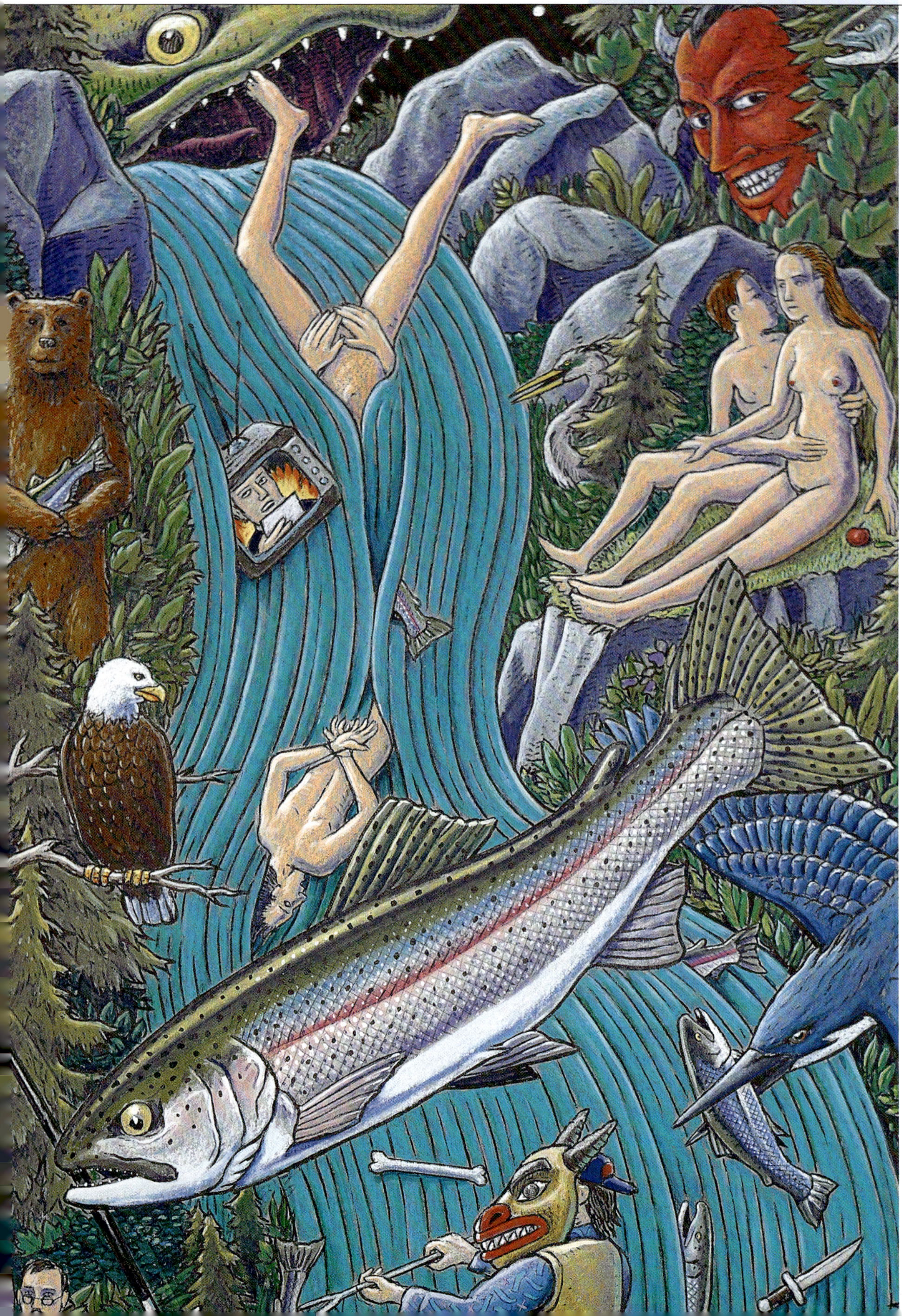

The titular pun that serves as the source for this image immediately brought lots of evocative images to mind. There's a struggle between good and evil going on here (note the grinning devil and the all-American eagle). Sexual overtones abound, shamelessly lifted from Bosch's *Garden of Earthly Delights*. There's a man pierced by a trout in the lower left, and Sigmund Freud himself can be seen scribbling notes on him and his condition. My wife, Michelle, can be seen just behind Freud, looking slightly bemused. There's a phrenology mannequin in midstream and a Tibetan devil dancer netting a fish. And of course there is the ever-questing fisherman at the bottom, being led deeper into his psyche by the elusive rainbow trout.

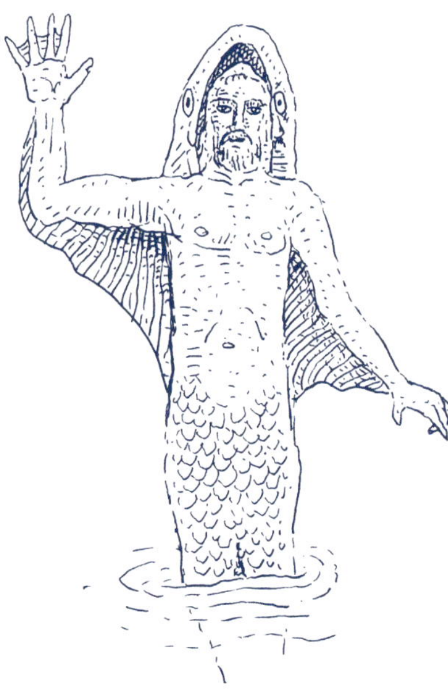

Detail, top

1991, mixed media on paper, 34 in. × 15 in.

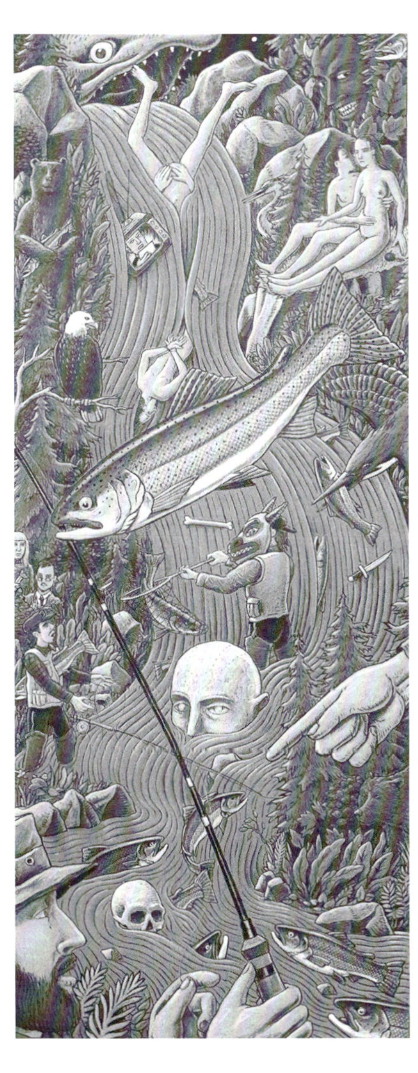

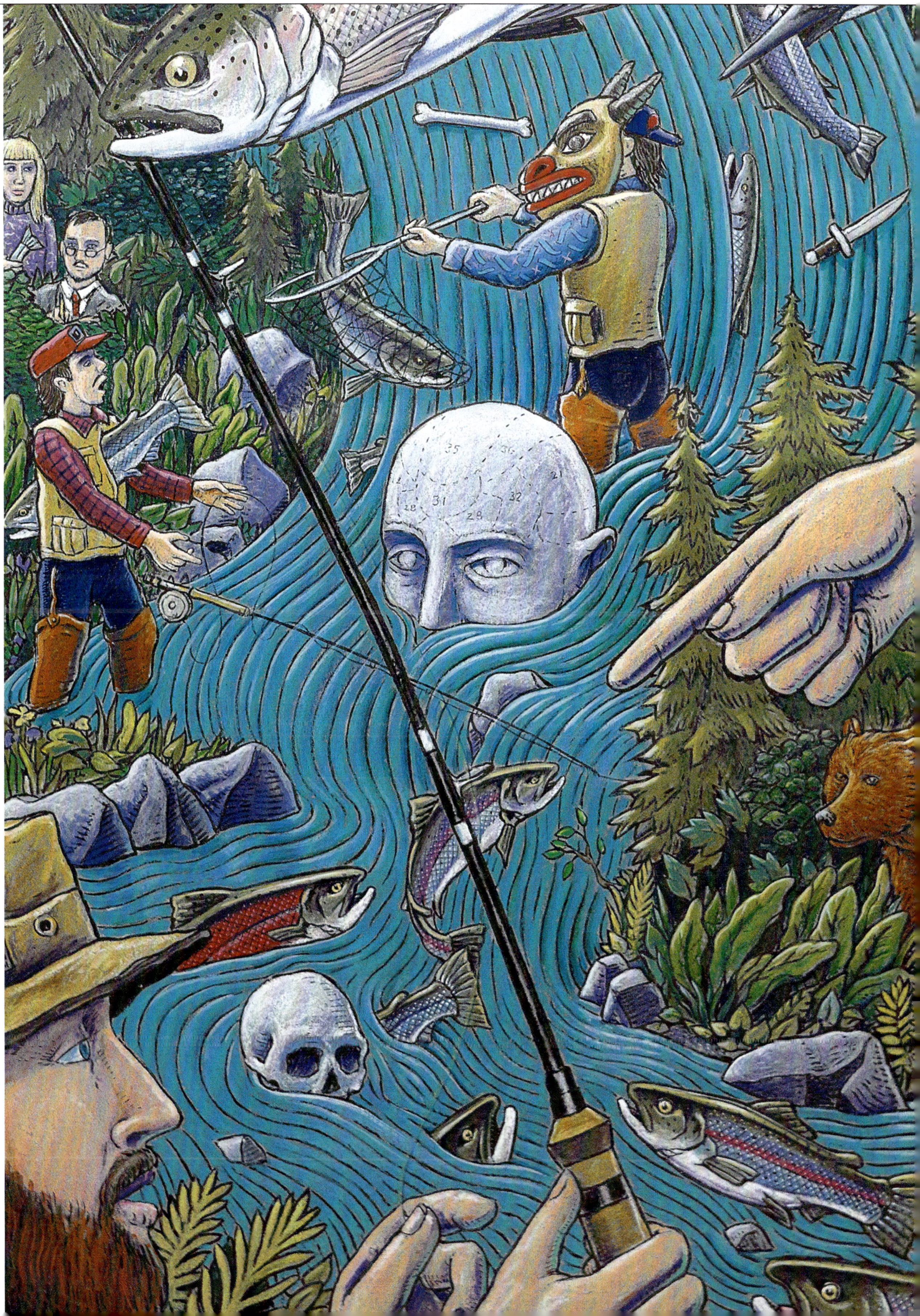

Detail, bottom

RAPTURE OF THE DEEP

SCUBA DIVERS who go a little too deep sometimes report an odd dreamlike sensation called "rapture of the deep" or the "martini effect." It's caused by too much nitrogen building up in the blood, and it can cause divers to hallucinate. I'm only an ardent snorkeler, so I've never experienced it, but this is my artistic interpretation of the experience. The dazed-looking helmeted diver in the lower right is surrounded by a whole menagerie of my deep-water favorites, from ratfish to coelacanths.

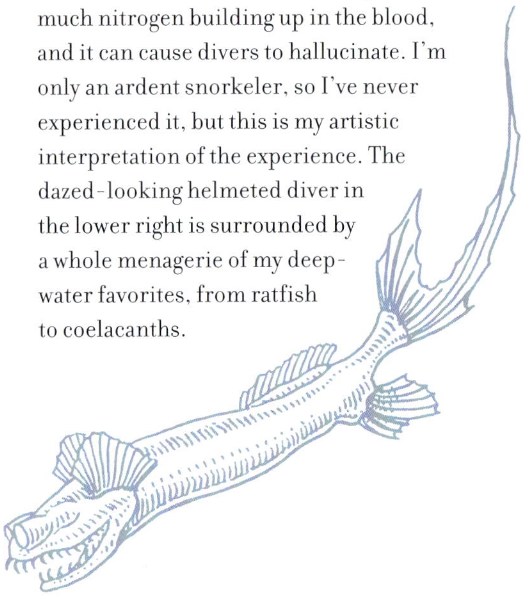

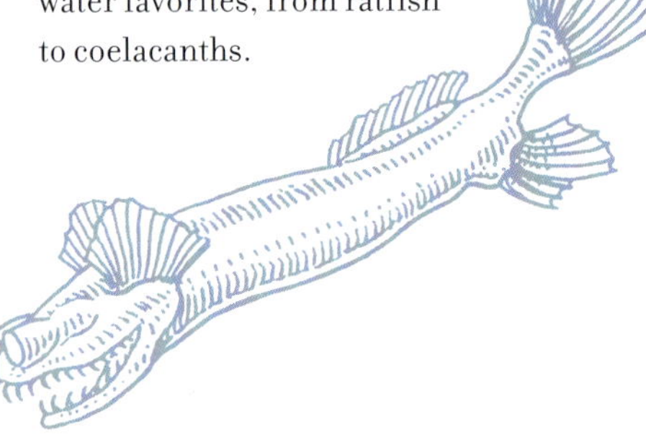

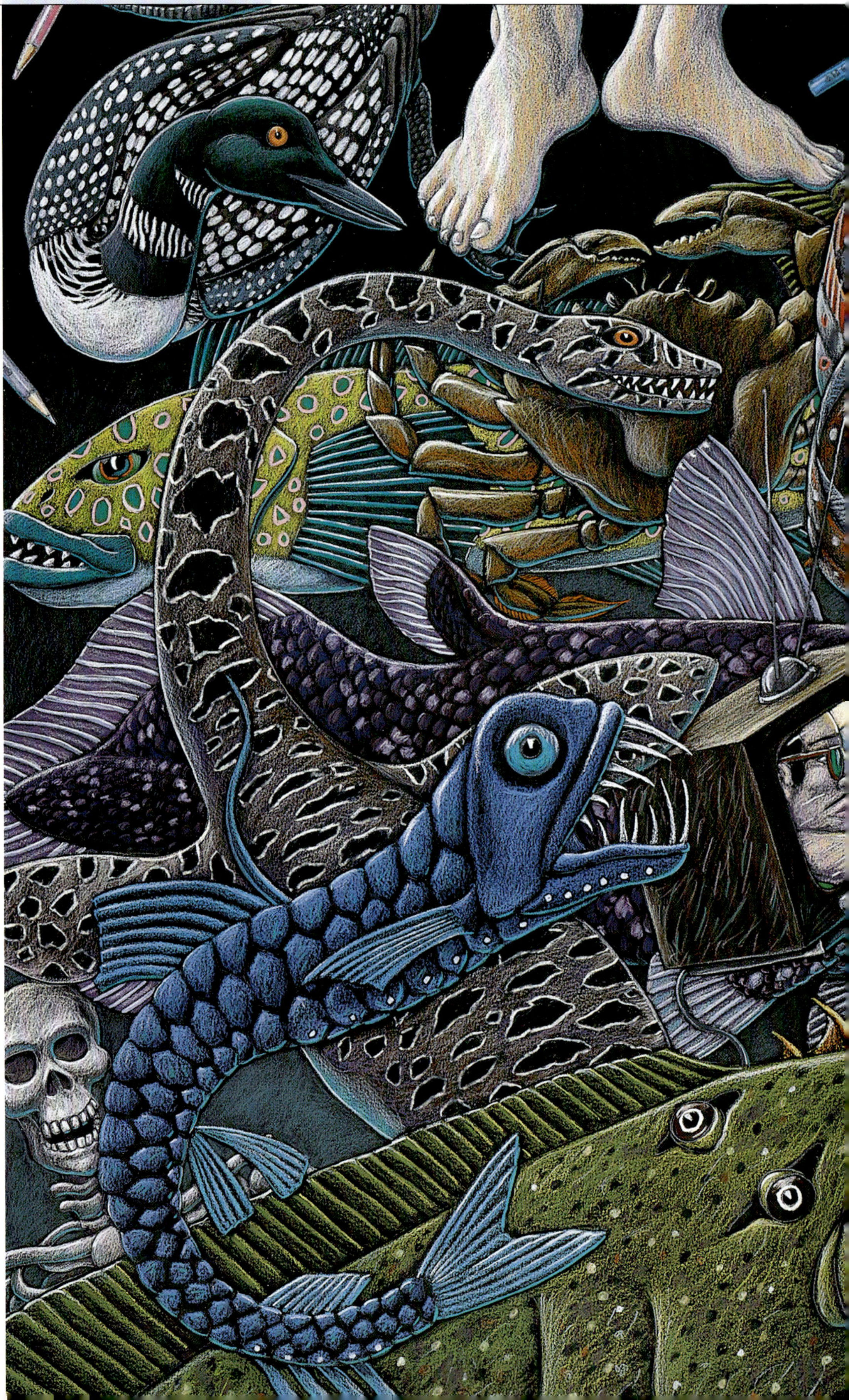

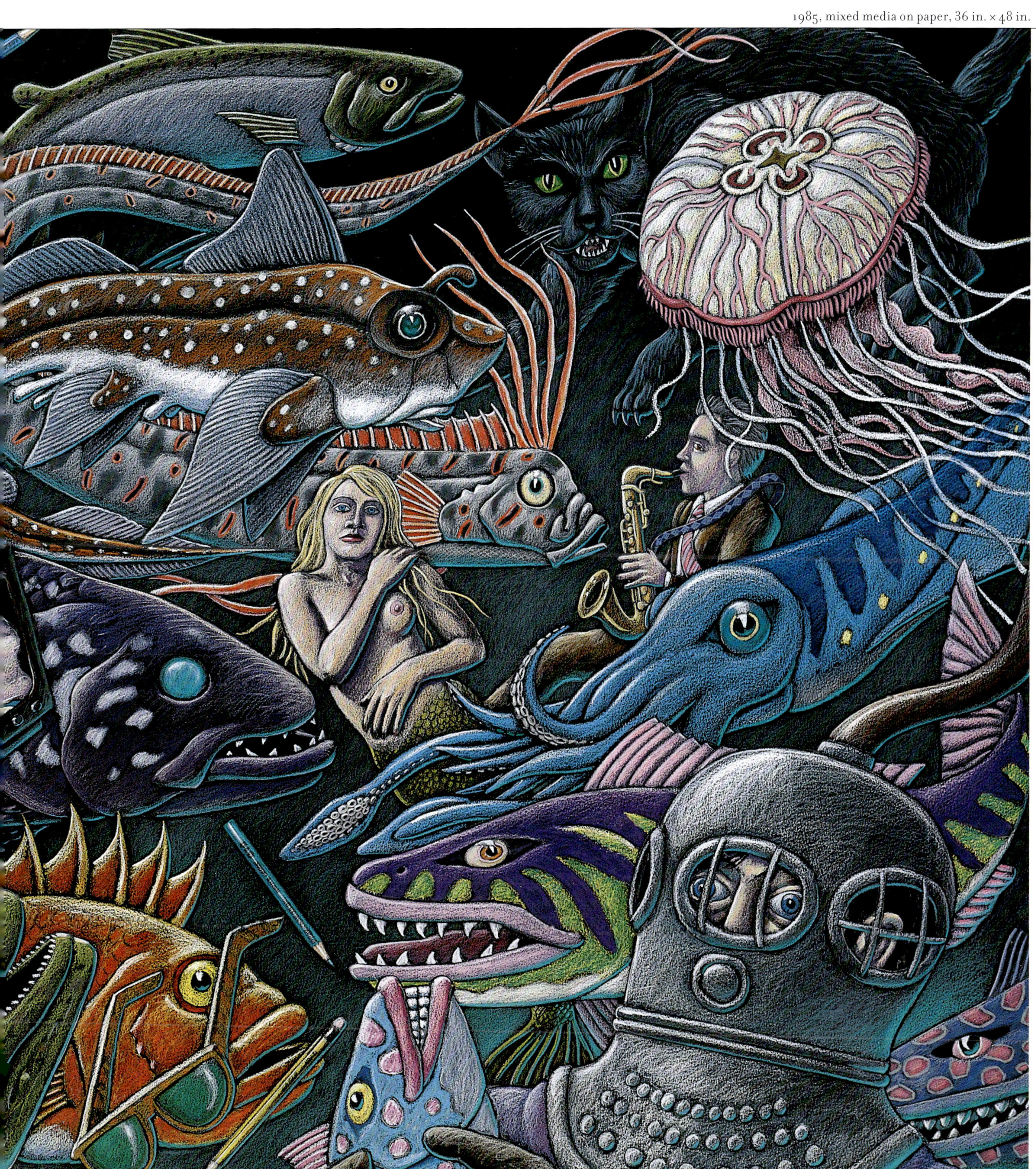

1985, mixed media on paper, 36 in. × 48 in.

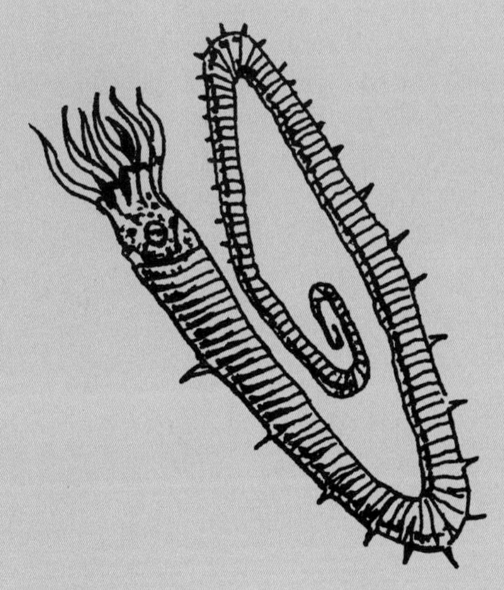

DANCING TO THE FOSSIL RECORD

1998, pen, ink, and watercolor on paper, 7 in. × 5.5 in.

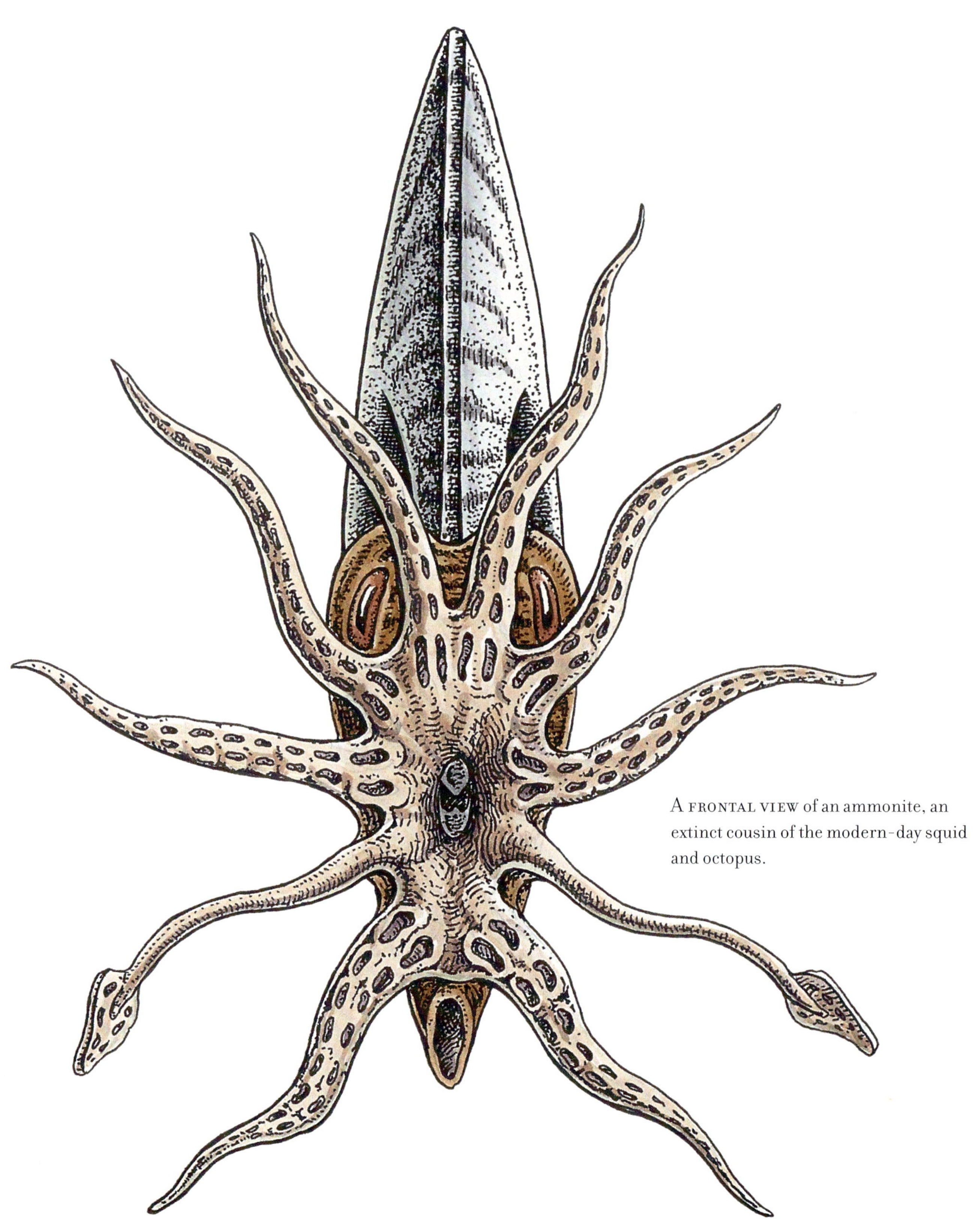

A FRONTAL VIEW of an ammonite, an extinct cousin of the modern-day squid and octopus.

PROFESSOR WILLISTON'S LAST SUPPER

1982, colored pencil on paper, 20 in. × 28 in.

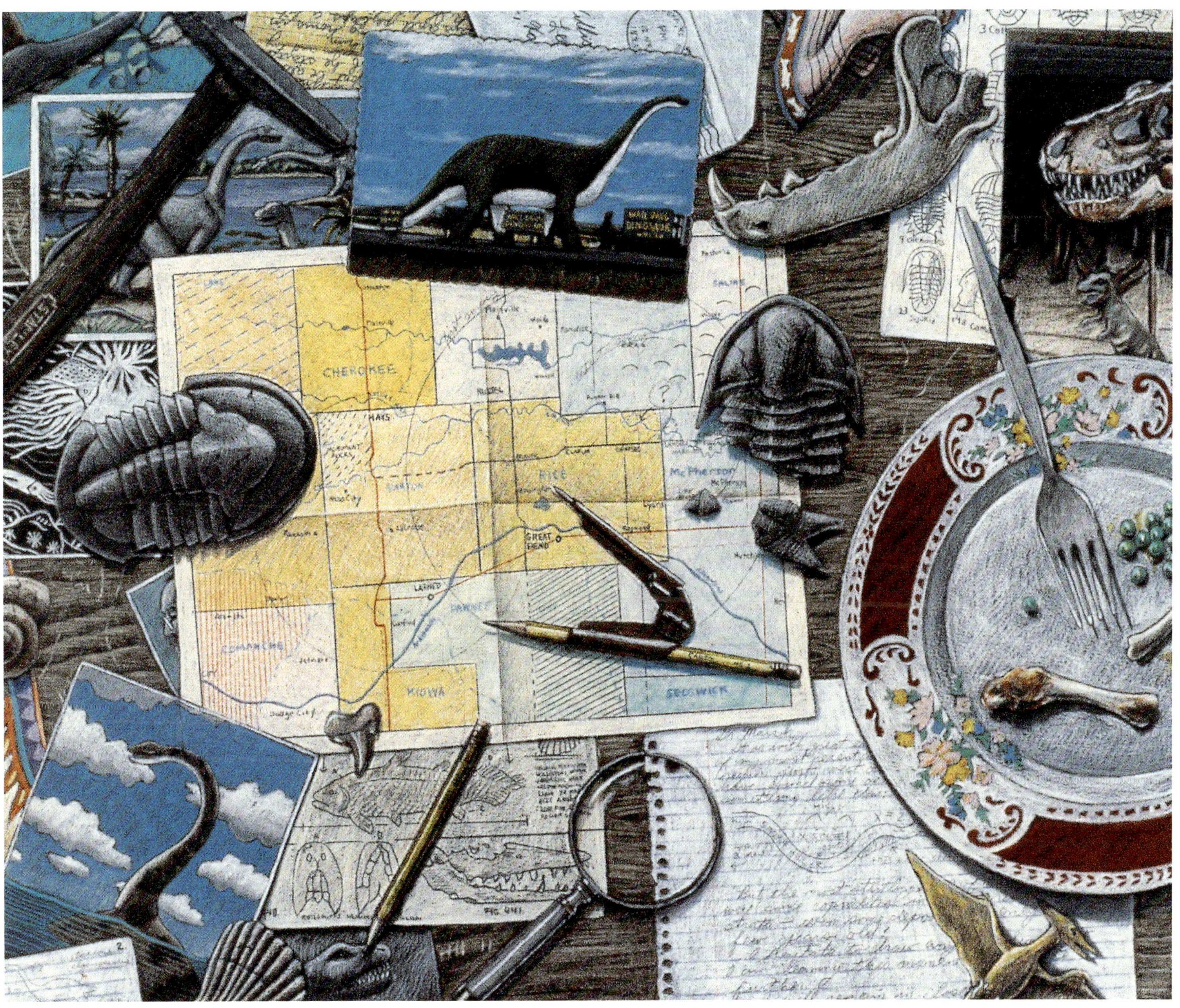

DR. SAMUEL WILLISTON was a paleontologist who taught at the University of Kansas in the late 1800s. He was an expert on the marine reptiles that flourished in that part of the world eighty million years ago, when Kansas was covered by a vast inland sea. That's him on the left, barely poking his nose out from under the map beneath the trilobite. This is an imaginary still-life interpretation of what his cluttered desktop might have looked like. He's just received a letter about a newly discovered fossil site and has become so excited that he left in a flurry and hasn't finished his dinner of chicken and peas.

TRILOS BY FIRE

1995, pastel on paper, 19.5 in. × 25 in.

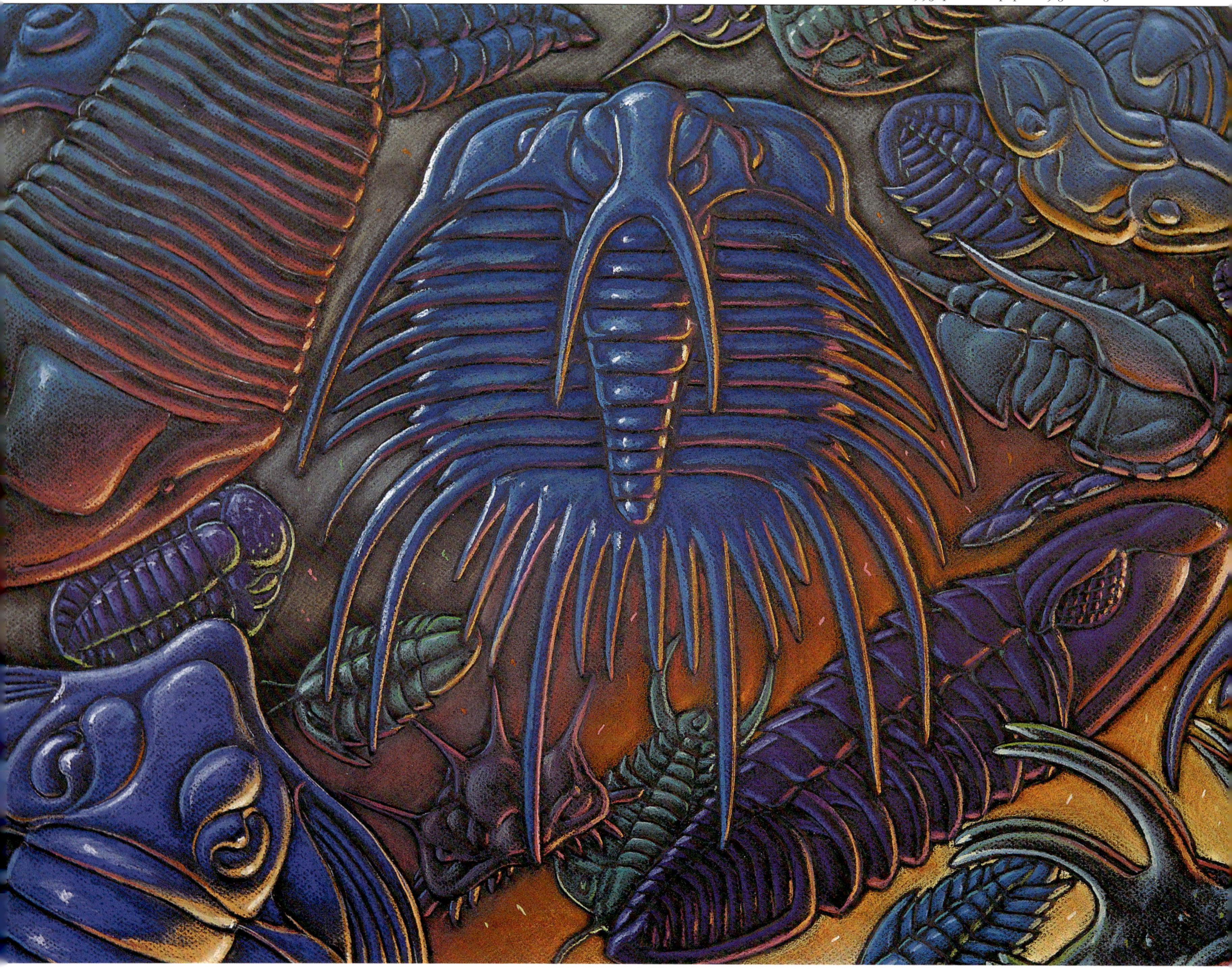

TRILOBITES ARE an extinct class of segmented marine animals that thrived in the oceans for nearly three hundred million years before they vanished forever at the end of the Permian period. With over fifteen thousand described species, their diversity is overwhelming. I've never tired of their elegant forms. They look so ancient yet have a weird kind of *Star Wars* futuristic appearance too.

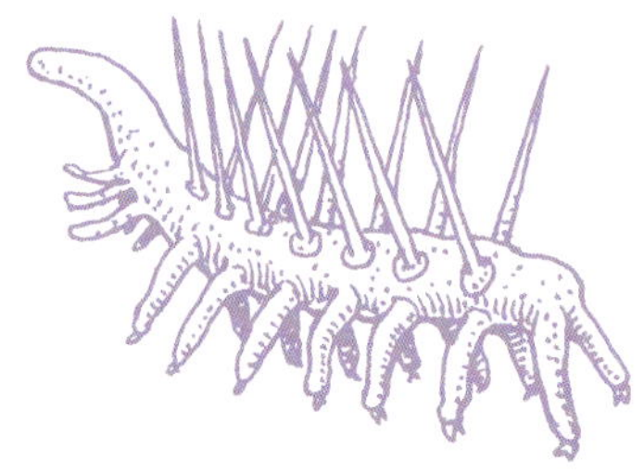

1995, mixed media on paper, 19 in. × 25 in.

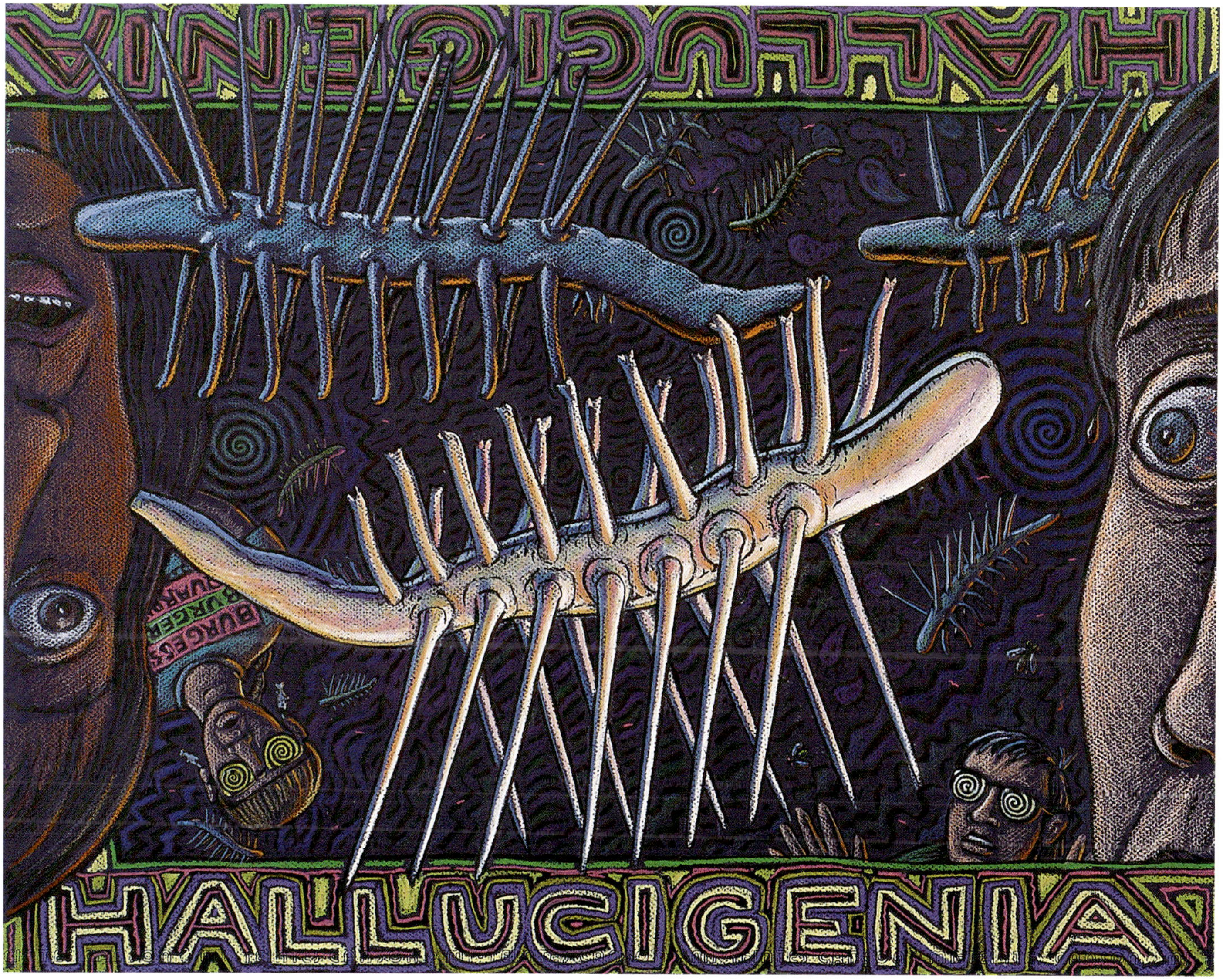

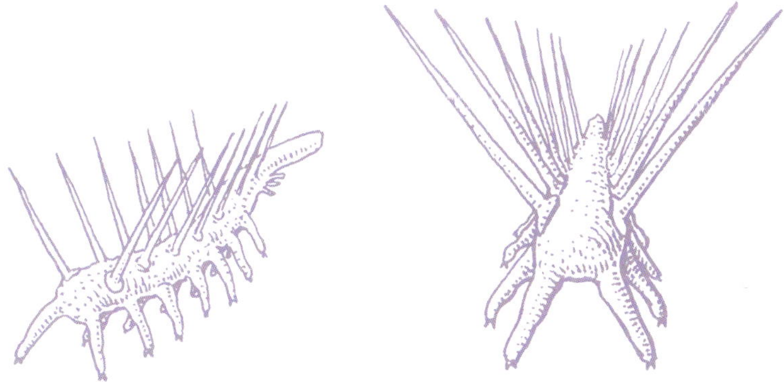

Hallucigenia, an extinct creature from the Cambrian period, looked so strange that it's easy to see why scientists call it what they do. It's so weird, in fact, that it actually took researchers a decade before they realized they were looking at the fossil upside down. Just to cover my bases, I've made this drawing work if you stand on your head as well.

THE LUCKY FISH GETS THE CHEESEBURGER

1995, pastel on paper, 25 in. × 19.5 in.

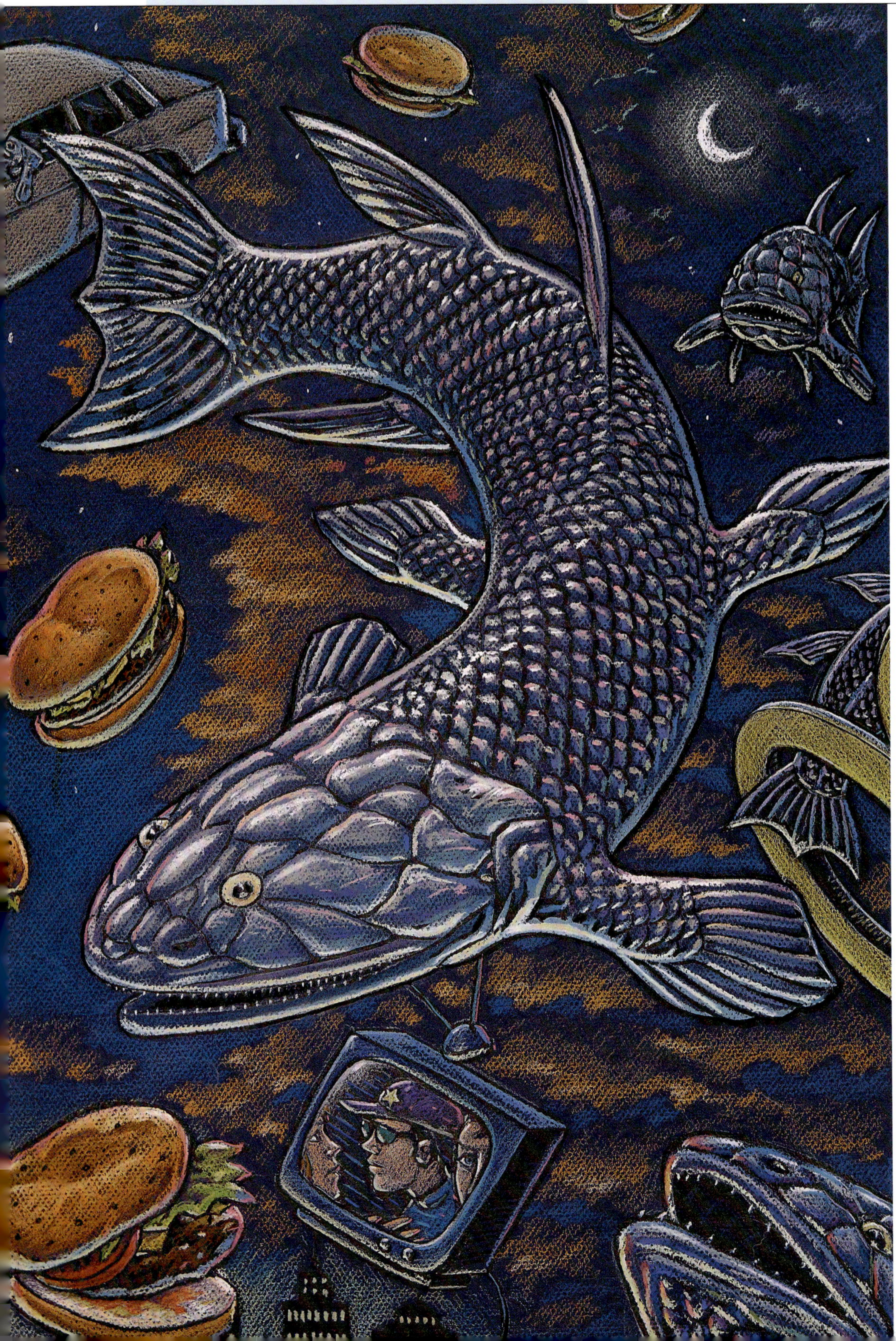

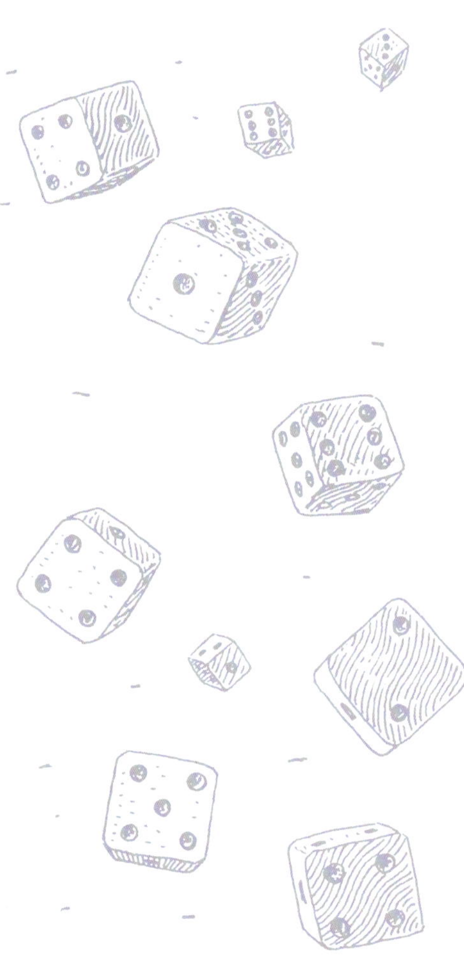

When Brad and I first experienced the "we *are* fish" epiphany while working on *Planet Ocean*, we thought it was hilarious. What if the lobe-finned fish left the water all those millions of years ago just to evolve into humans and experience the joys of eating cheap cheeseburgers, watching bad TV, and driving fast cars? Note the golden arches on the right. This Devonian-period fish is called *Eusthenopteron*, meaning "strong fin," a reference to the limblike bones in its fins.

IN THE MIDDAY SUN . . .

1999, acrylic on board, 15 in. × 12 in.

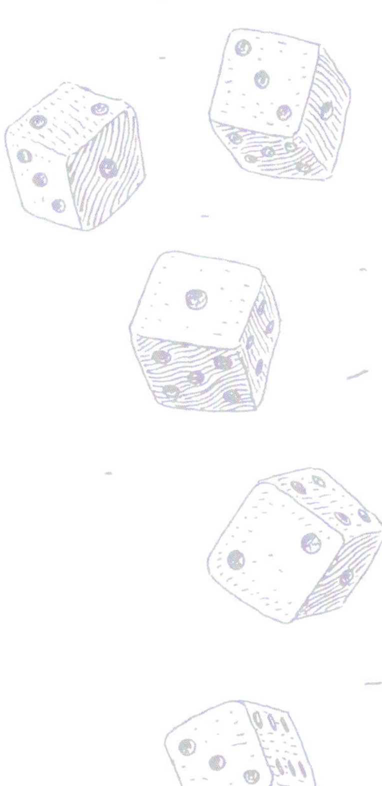

IN THE MIDDAY SUN I IMAGINED

AMMONITES SWIMMING IN THE DESERT SKY

In the summer of 1998 I joined Dr. Kirk Johnson and a crew of volunteers from the Denver Museum of Nature and Science in excavating Cretaceous-period ammonites from a hillside near Kremmling, Colorado. As I dug for these buried beasts, the sun beating down on my pale Alaskan skin, I began to daydream . . .

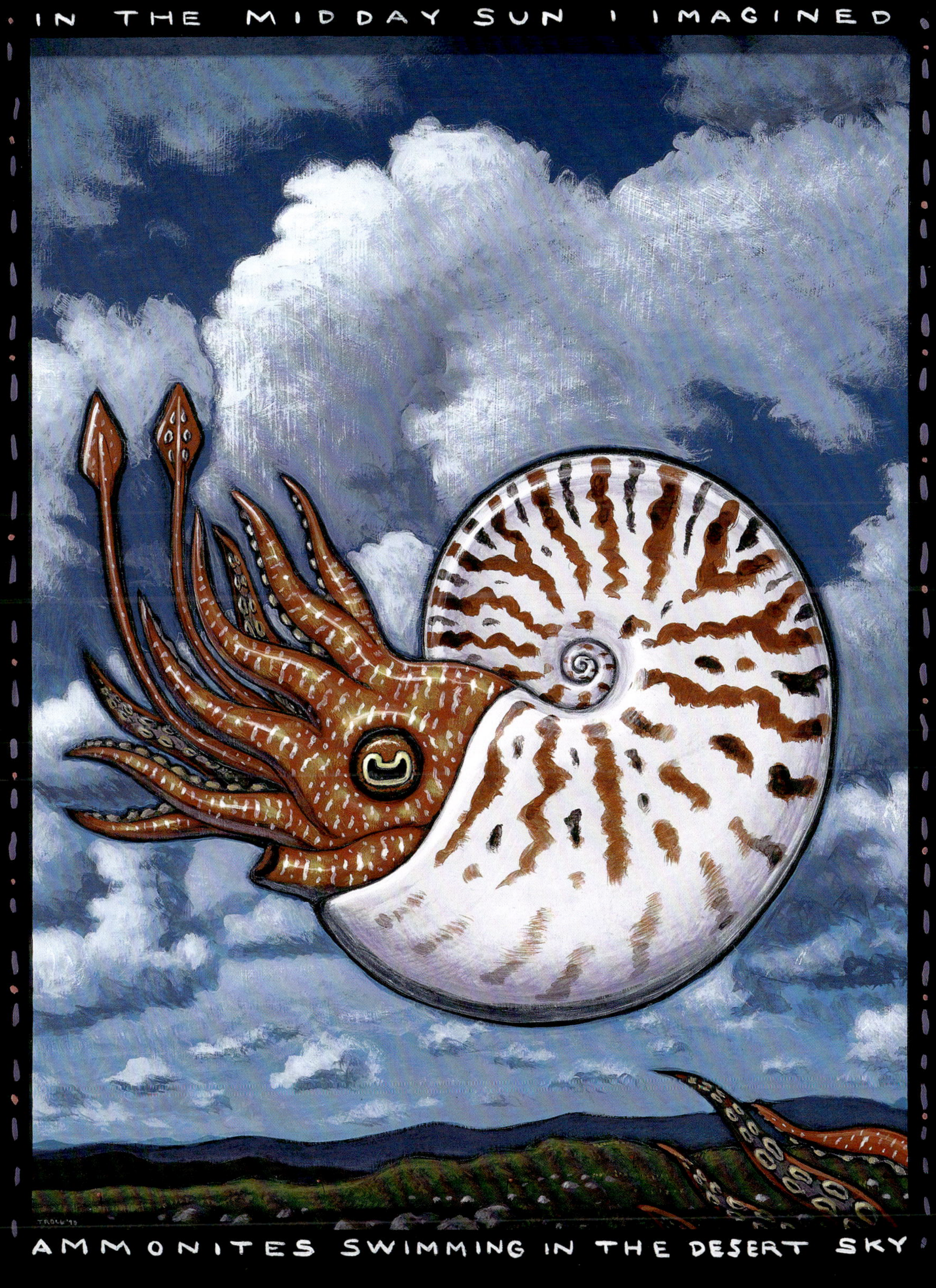

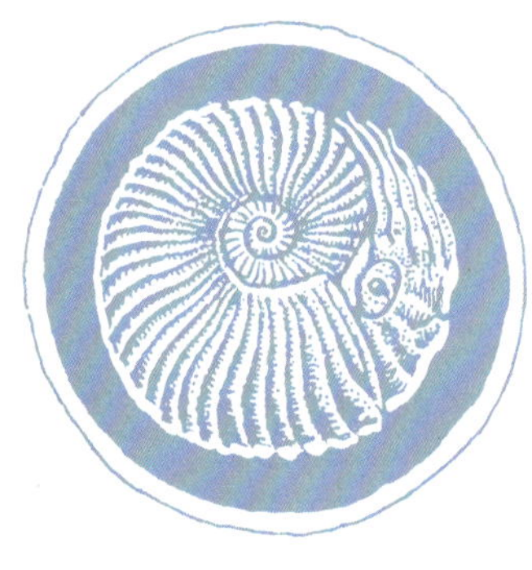

Ancient ammonites came together in large aggregations to spawn and die, much as modern-day salmon do. The females of this genus, *Placenticerus*, reached the size of automobile tires. It's likely that they timed their rendezvous with the much smaller males using a lunar cycle, hence the moonlit night. Avert your eyes if you must, but you can actually see tentacle-to-tentacle ammonite sex taking place in the lower center. Marine reptiles are darting in to snarf down a few of the preoccupied couples.

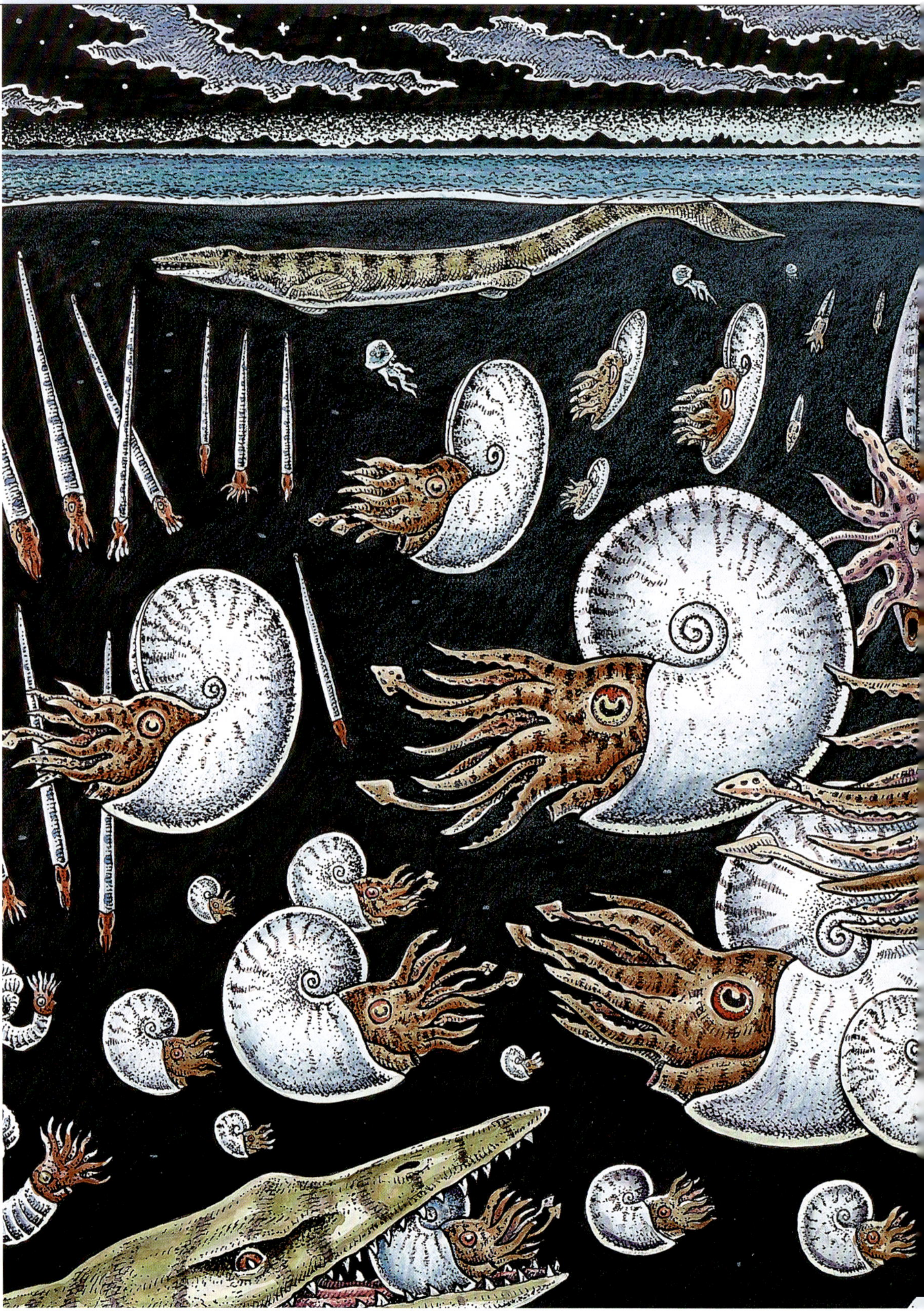

A MOONLIT NIGHT SOMEWHERE NEAR THE

1998, pen, ink, and watercolor on paper, 11 in. × 22 in.

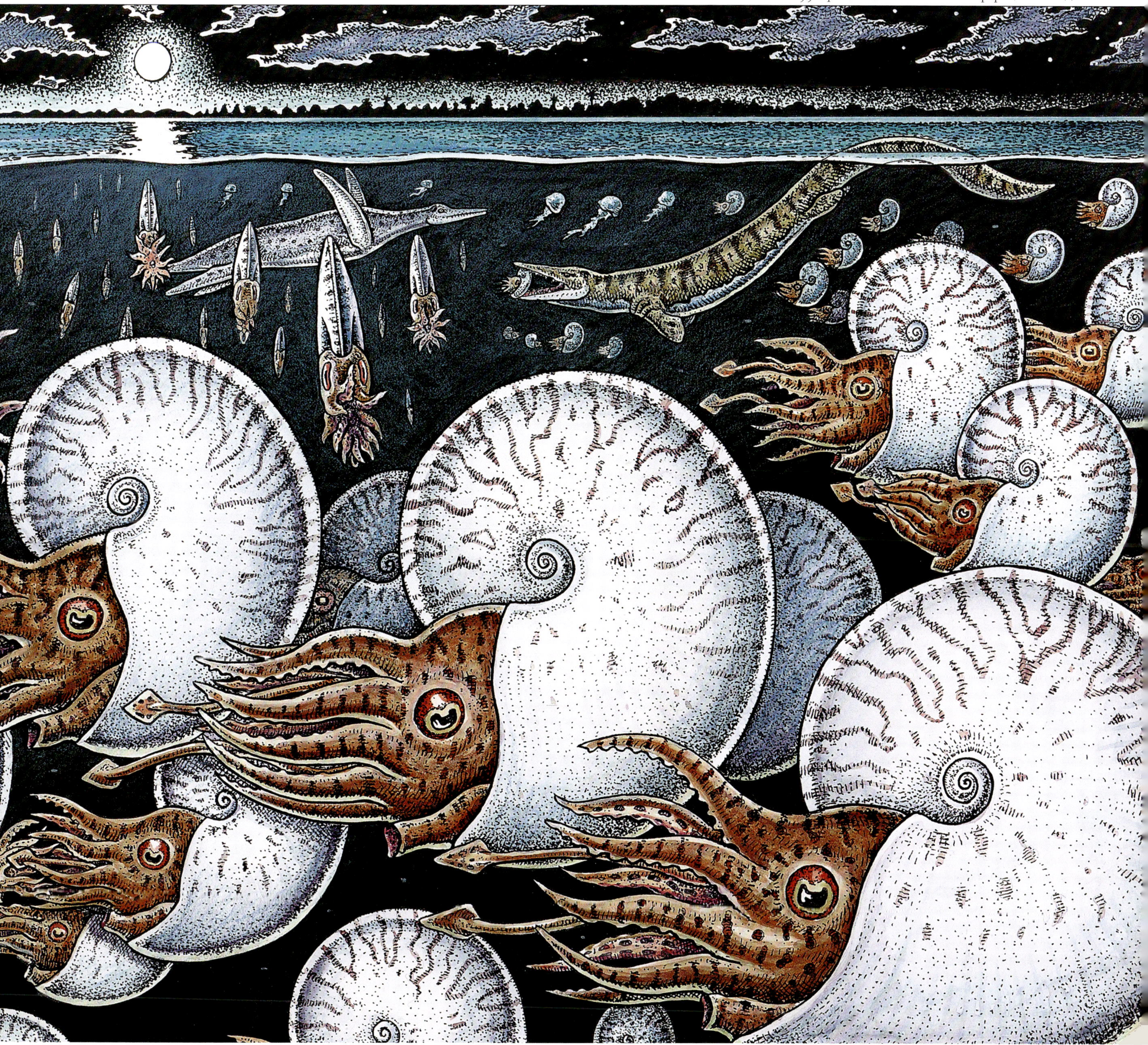

TOWN OF KREMMLING, COLORADO SEVENTY THREE MILLION YEARS AGO

1995, pastel on paper, 19.5 in. × 25 in.

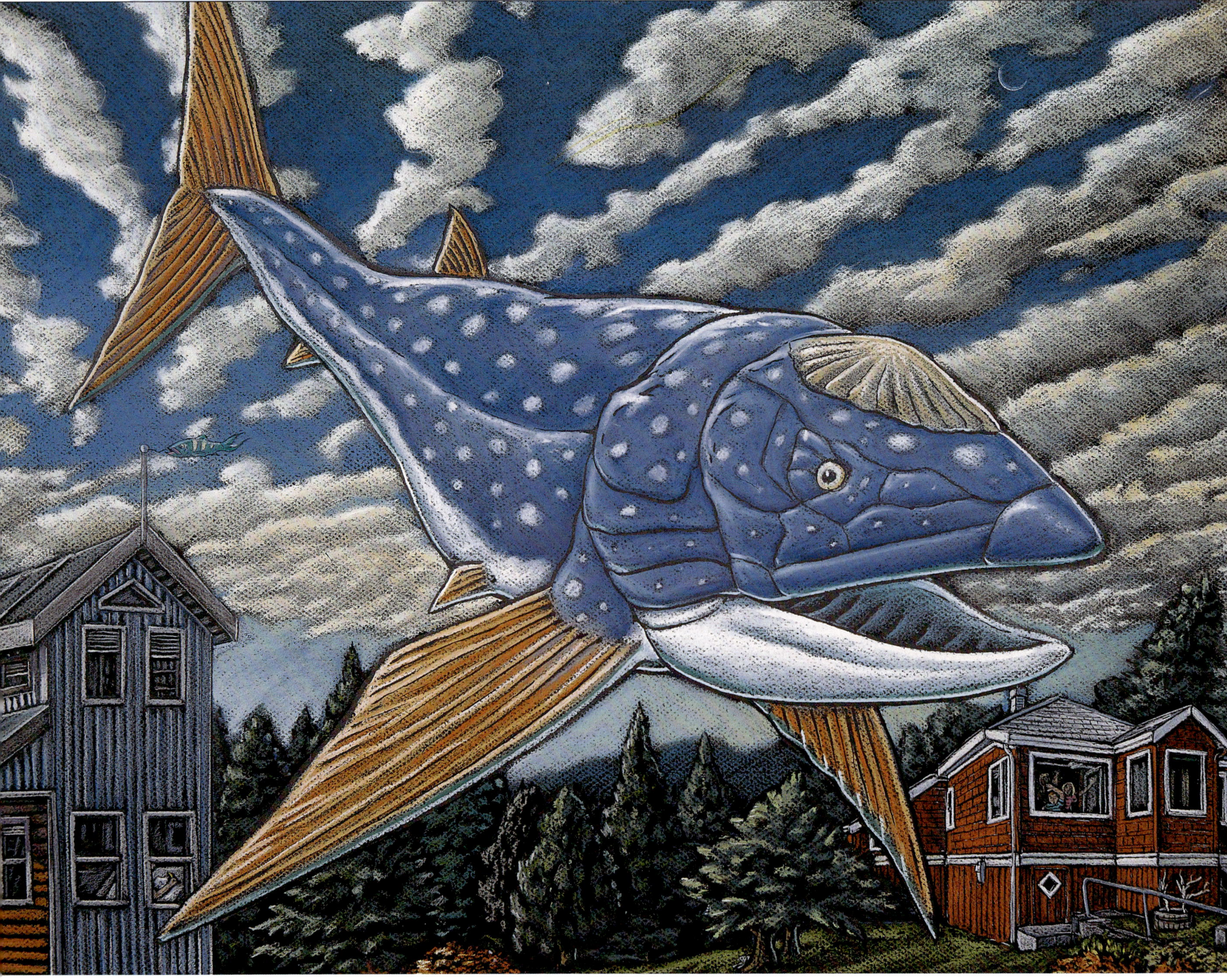

This is a Jurassic-period bony fish from England that may have reached a length of one hundred feet, making it the biggest fish of all time. That's my house on the right, with my family in the window. Our cat can be seen peeking out from behind the tree at center. I'm out working in my studio, there on the left. The goofy-looking thing on the fish's head is a bony plate, not a hat.

TRILOBITE SAFARI

1993, colored pencil on paper, 5.5 in. × 8.25 in.

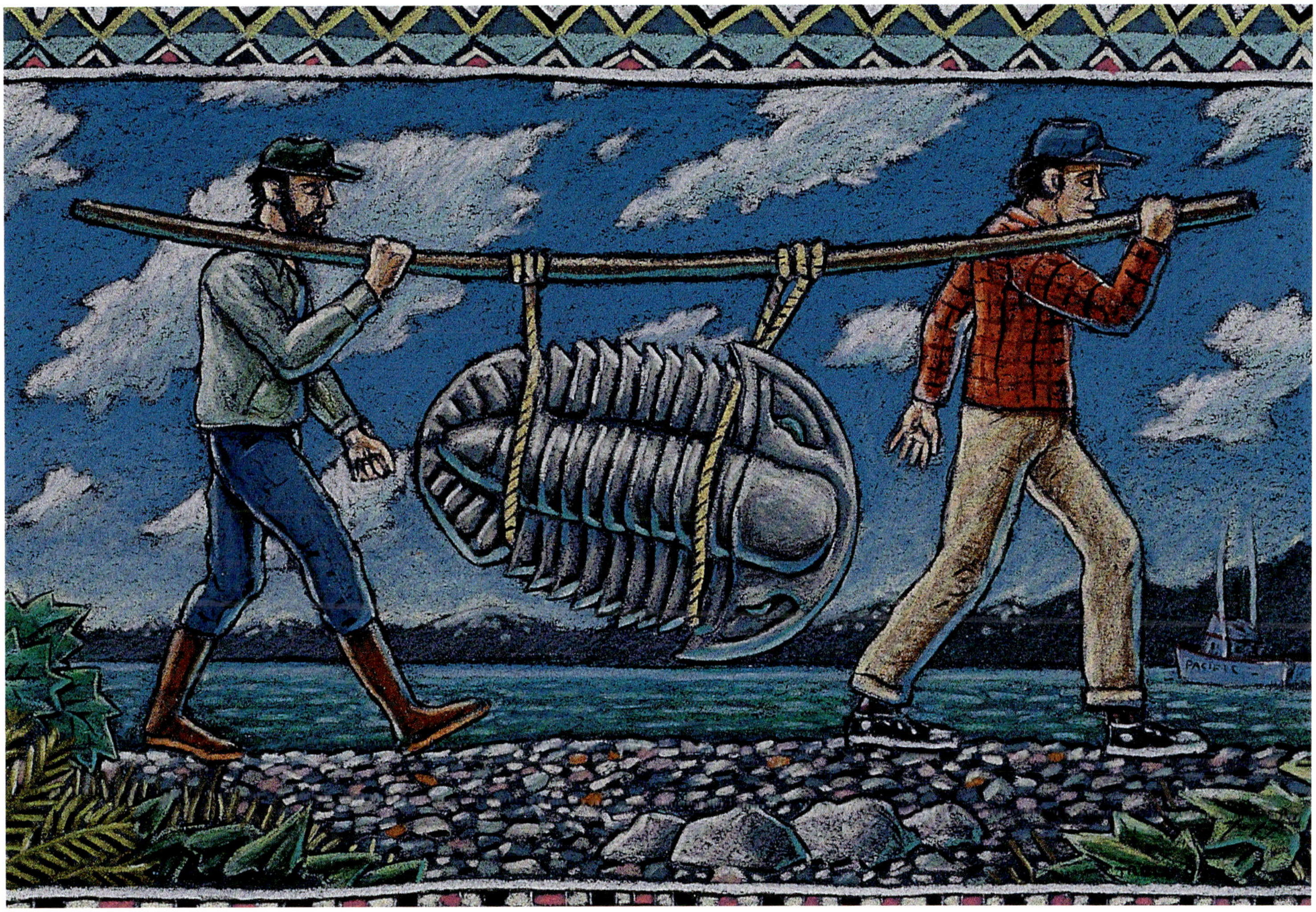

AH, IF ONLY TRILOBITES reached this size! But alas, the biggest one ever found by anyone is only about two feet long. It's hard to describe the pure joy I felt when I finally bagged my very first trilobite on an island in Southeast Alaska. It's only an inch long, but it's priceless to me.

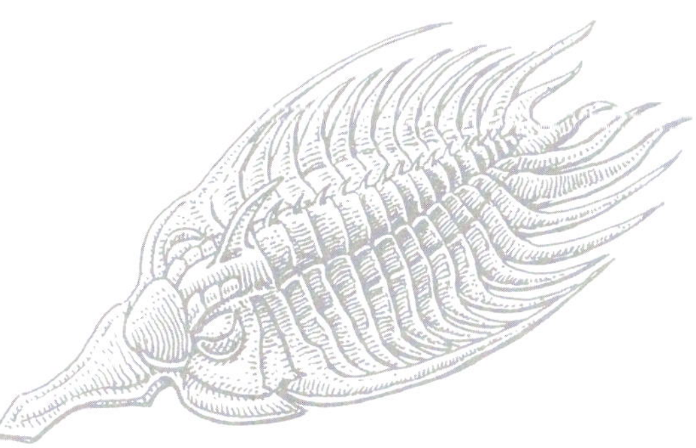

1993, pastel on paper, 18 in. × 22 in.

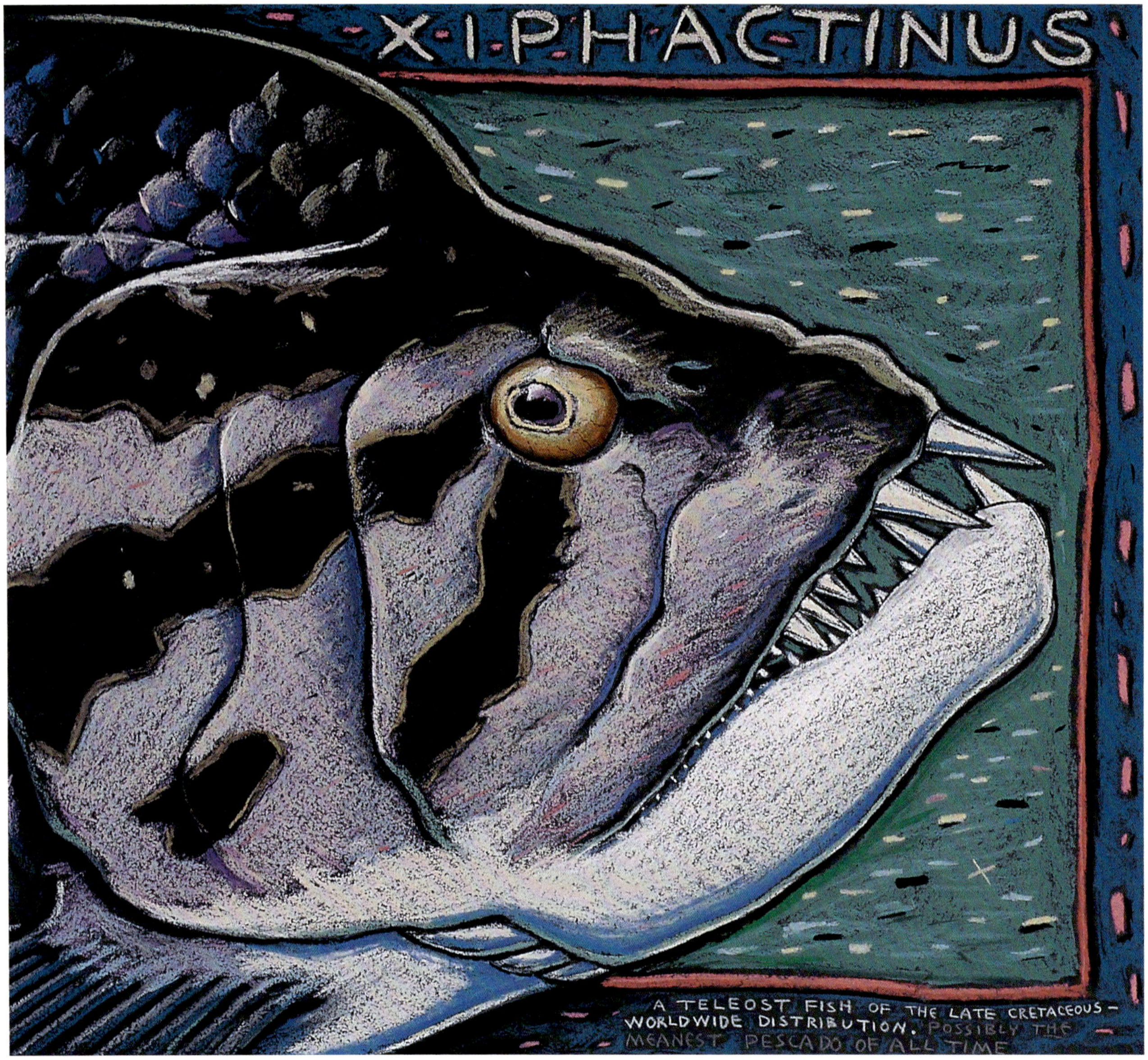

THIS CANTANKEROUS-LOOKING fish is called *Xiphactinus*, which means "sword ray." Its fossils have been found all over the world, but the best ones come from my former home state of Kansas. They averaged about fifteen feet in length and must have been the terrors of the Mesozoic seas with their gigantic teeth reinforced by massive bony jaws. I like to call them the "meanest pescados of all time."

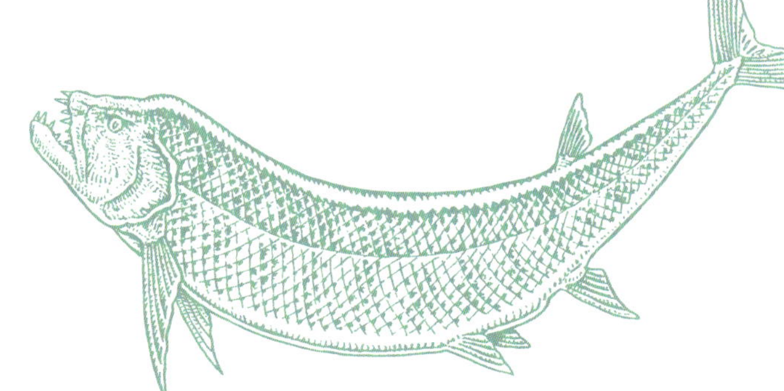

CRETACEOUS ROAD DREAM

1992, colored pencil on paper, 10 in. × 7.5 in.

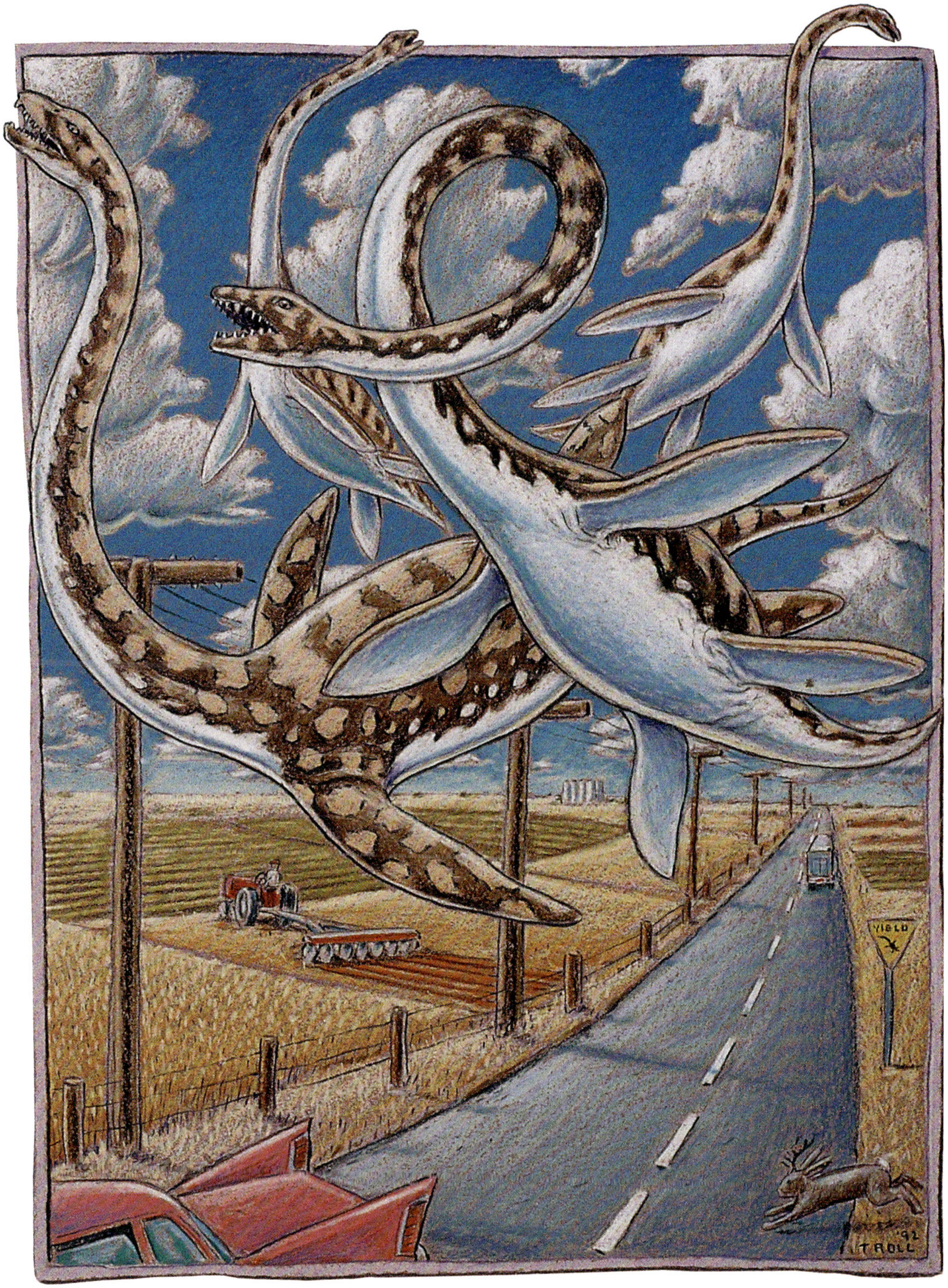

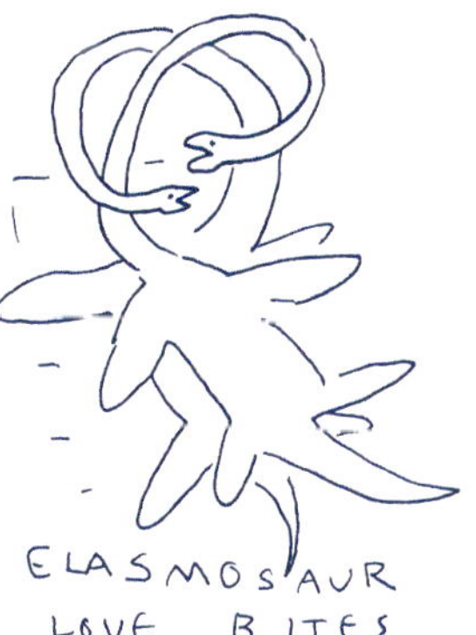

IT'S EASY TO IMAGINE the rolling wheat-covered hills and plains of Kansas as an ancient sea. Speeding along the straight and narrow back roads with the windows rolled down on a hot summer day, recovering from the previous night's excess, I saw serpentine plesiosaurs like these gliding through the cumulus clouds.

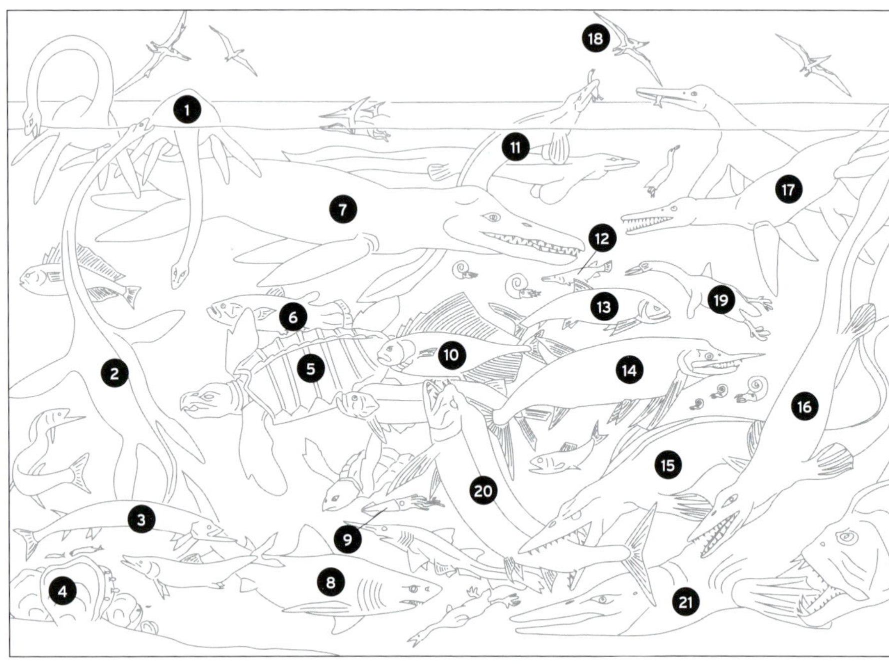

An inland sea once divided the North American continent during the Cretaceous period, stretching from Texas all the way through Canada. It was home to a vast array of creatures, from enormous seagoing reptiles right on down to goofy-looking diving birds. Some of the choicest fossils from this lost world come from the chalk rocks of western Kansas. The only animal in this picture that is still alive is the lowly gar, which you can see right below the big-headed plesiosaur at the top center. Although it's a freshwater fish, its scales have been found in the chalky marine deposits known as the Niobrara formation. Further proof that wherever you go, there you gar.

1. *Styxosaurus* (plesiosaur)
2. *Thalassomedon* (plesiosaur)
3. *Saurodon* (bony fish)
4. *Platyceramus* with *Omosoma* (clam and schooling fish)
5. *Archelon* (marine turtle)
6. Giant coelacanth
7. *Brachauchenius* (short-necked plesiosaur)
8. *Cretoxyrhina* (mako shark)
9. Squid
10. *Bananogmius* (bony fish)
11. *Clidastes* (mosasaur)
12. Gar
13. *Pachyrhizodus* (bony fish)
14. *Protosphyraena* (bony fish)
15. *Tylosaurus* (mosasaur)
16. *Platecarpus* (mosasaur)
17. *Trinacromerum* (short-necked plesiosaur)
18. *Pteranodon* (pterosaur)
19. *Hesperornis* (marine bird)
20. *Xiphactinus* (bony fish)
21. *Hainosaurus* (mosasaur)

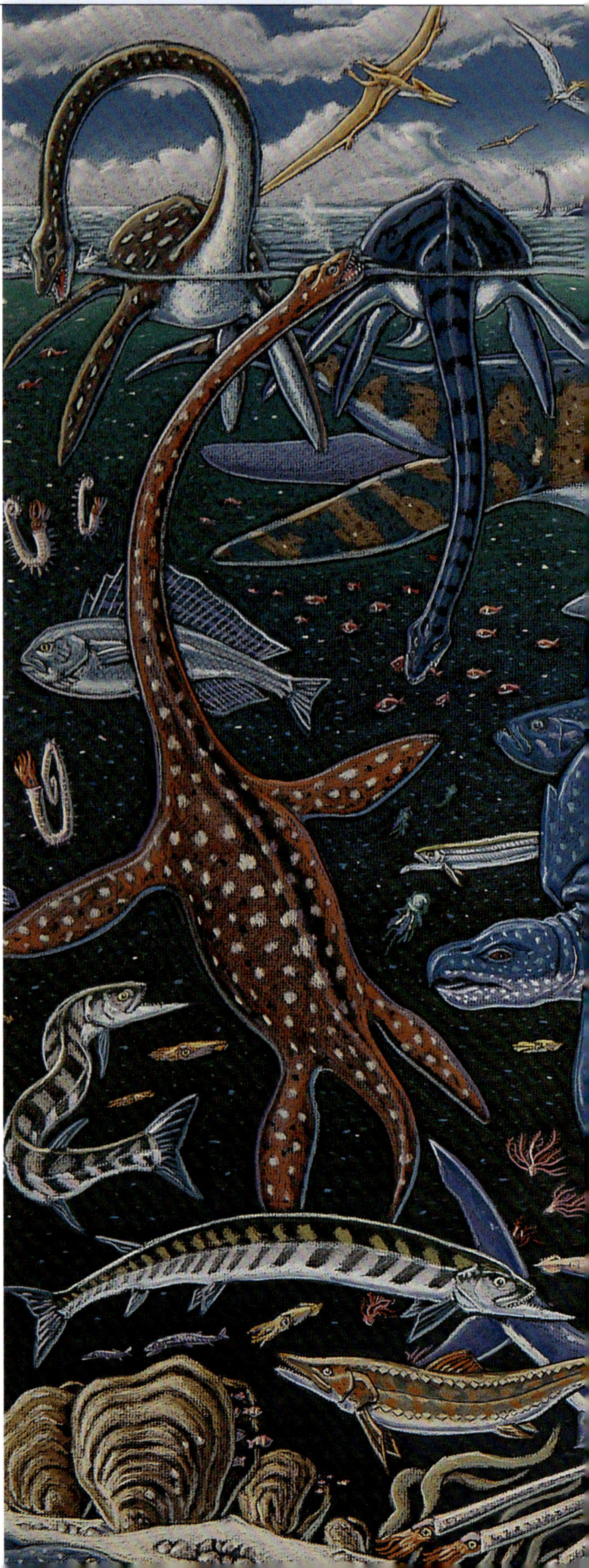

1993, pastel on paper, 36 in. × 48 in.

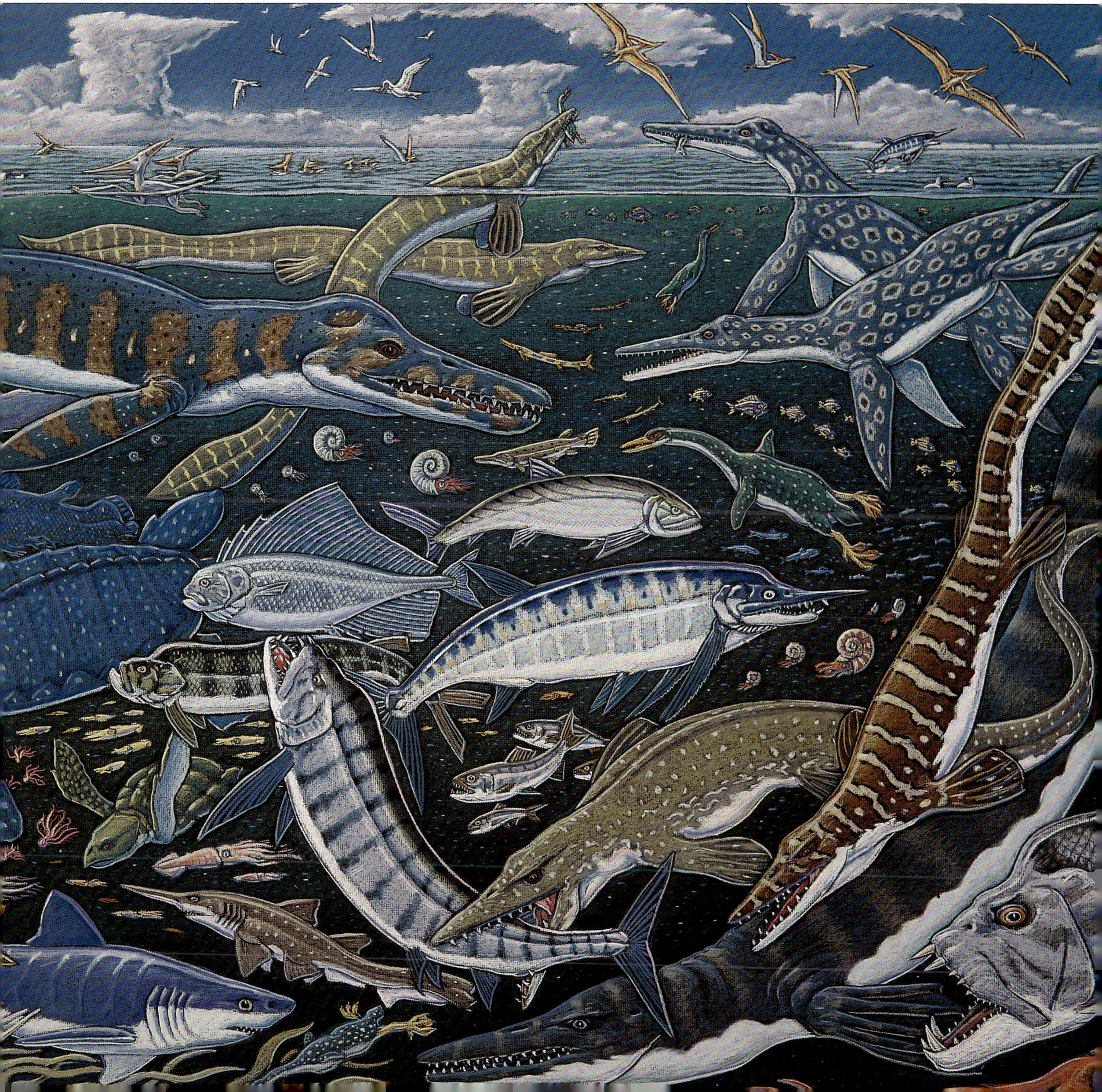

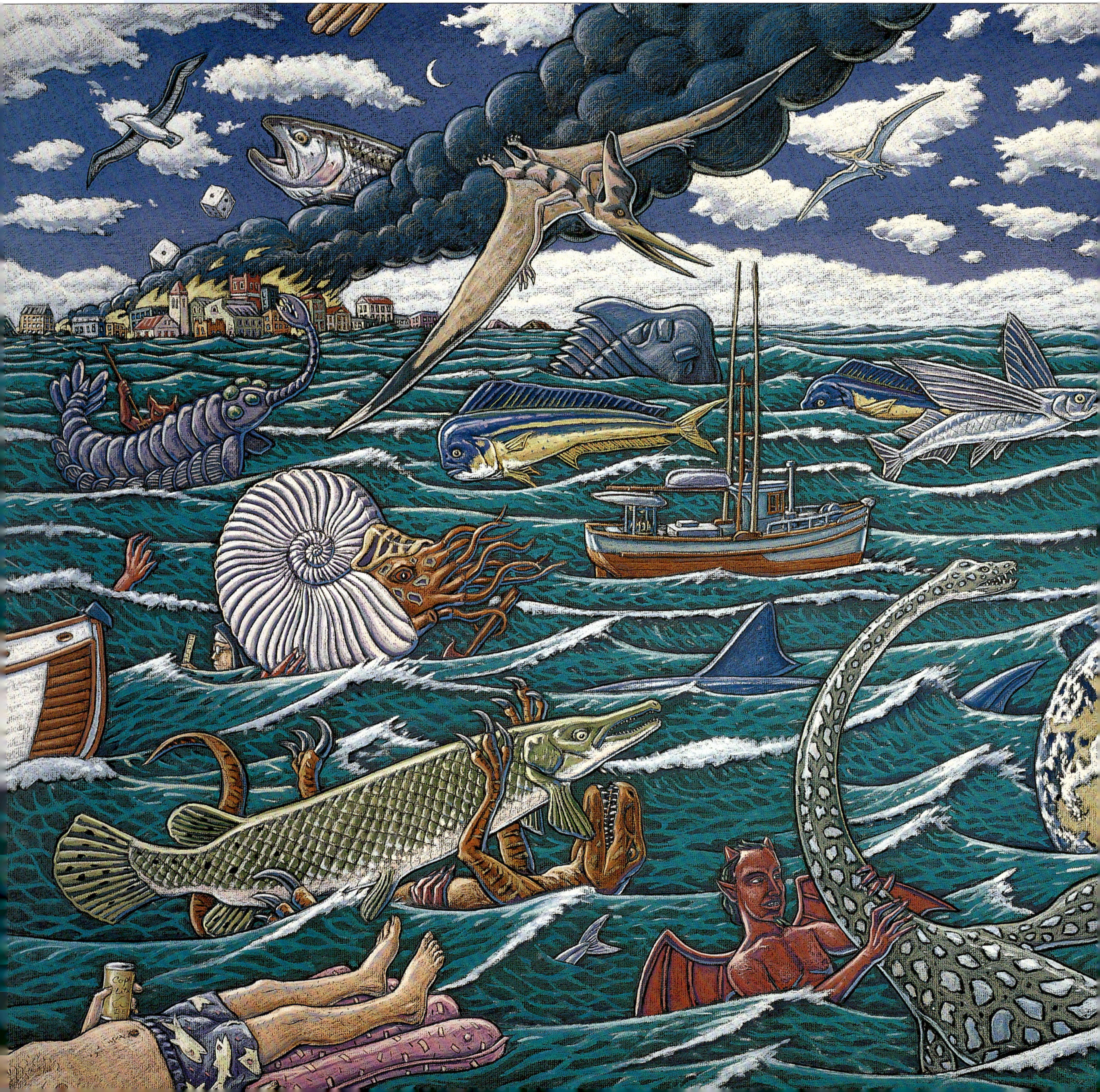

1994, pastel on paper, 36 in. × 48 in.

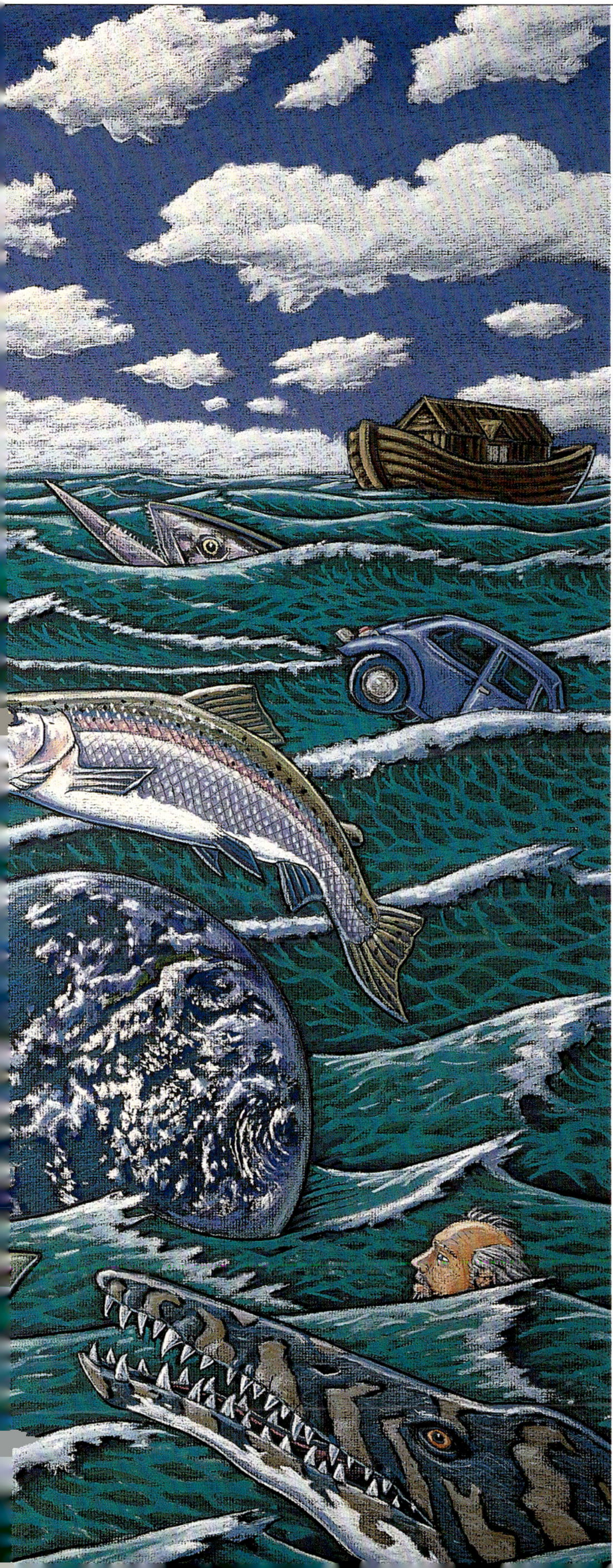

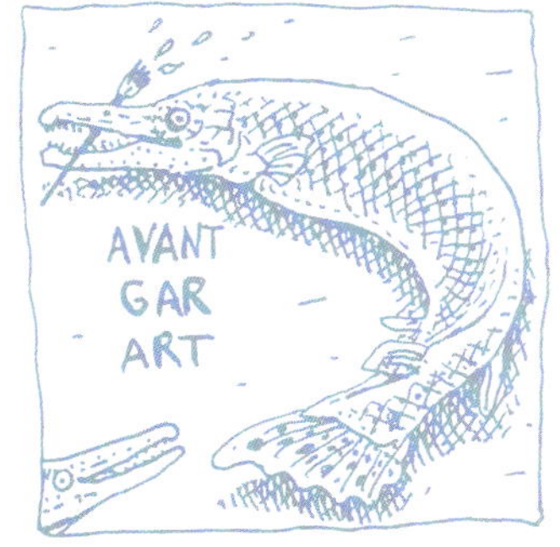

In 1994, Brad Matsen and I collaborated on the book *Planet Ocean*, which explores the evolution of life on our planet from its watery beginnings in the sea. The narrative and art are built around a road trip Brad and I took from Alaska to Kansas in 1992. For the book cover, I did this large pastel drawing that incorporates elements from the narrative and hints at some of our discoveries and obsessions as we traveled through time by way of fossil beds and museums.

The automobile in the upper right corner is an emblem of our journey. The guy on the air mattress with the round belly in the lower left is yours truly. I'm holding a can of Copra Lite beer, a shameless pun on fossilized dung (coprolites). Brad's shiny head appears on the right, gazing at our planet adrift in high, rolling seas. Leaping over the globe is a magnificent steelhead, a reference to the fishing trip on an Alaskan stream that began our voyage of discovery.

The devil to Brad's left is grabbing the neck of an extinct marine reptile called a plesiosaur. When I was in second grade at Seven Sorrows School in Middletown, Pennsylvania, Sister Marie James told us all about Noah's ark (depicted in the upper right corner) and explained that dinosaurs were not allowed onboard as they were not part of God's plan. Being a precocious young paleo-nerd, I shot my hand up and asked, "But what about the plesiosaurs? They could swim and didn't have to be on the ark." I don't remember Sister Marie's response, but it was clear she had no idea what I was talking about. I sensed right then, at the tender age of seven, that there was a serious schism between the real history of life and the stories of religion. The ruler-brandishing Sister Marie appears on the left underneath the spiral ammonite, while the prow of Darwin's ship, *HMS Beagle*, can be seen to her left.

Elsewhere you can spot a gar wrestling with a velociraptor, the city of Los Angeles in flames, a pair of dice, the hand of God, an Alaskan salmon troller, an *Opibinia* from the Burgess Shale of Canada, a trilobite, a pterosaur, a flying fish, and a magnificent blue and yellow mahi mahi—all images that resonate with the themes of the book.

1997, linoleum block print and watercolor on paper, 19 in. × 18 in.

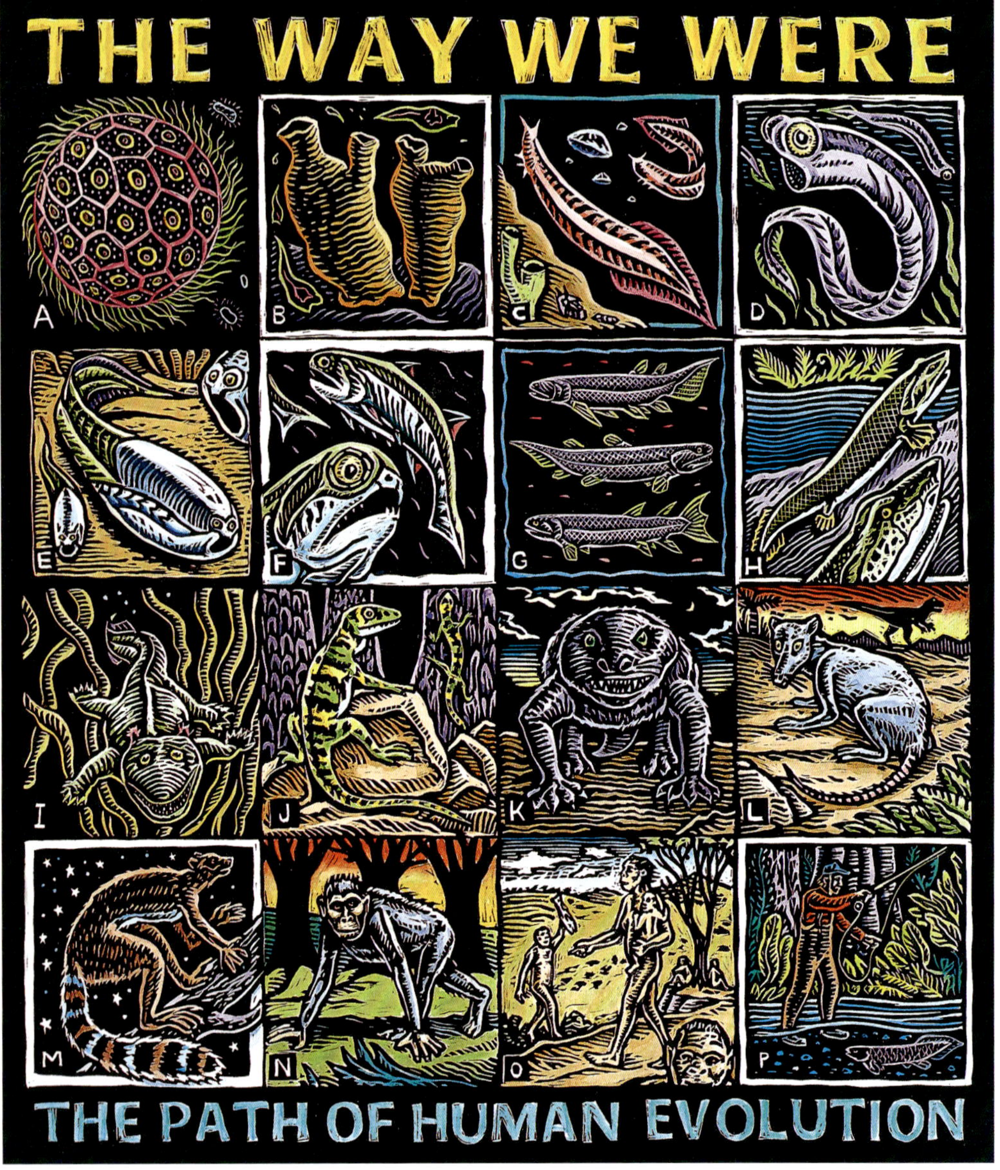

Most people understand the principle of evolution to some degree, and the average guy can see the biological connection between humans and the great apes. Few people know much about our lineage beyond that, though. The path that led to modern-day cheeseburger-chomping fly-fishing humans is a long and twisted one that ultimately connects us to the primordial soup of the ancient oceans. Tumbling backward through time, the procession of our ancestors reads like a freak show: small ratlike mammals, scaly lizardish creatures, funky-looking swamp-dwelling tetrapods (our Kermit phase), jazzy lobe-finned fishes, jawless fish that look like wayward vacuum cleaners, blind little wriggling wormlike animals with a hint of a backbone, and the ignoble spawn of spineless sea squirts.

An ichthyologist friend, Dr. Carl Ferraris, helped me work out the sixteen steps to humanity depicted here. I couldn't resist the idea of animating the sequence and, with the help of a few other folks, posted the results on my website (www.trollart.com).

FROM THE SLIME TO THE RIDICULOUS

1995, pen and ink on paper, 30 in. × 22 in.

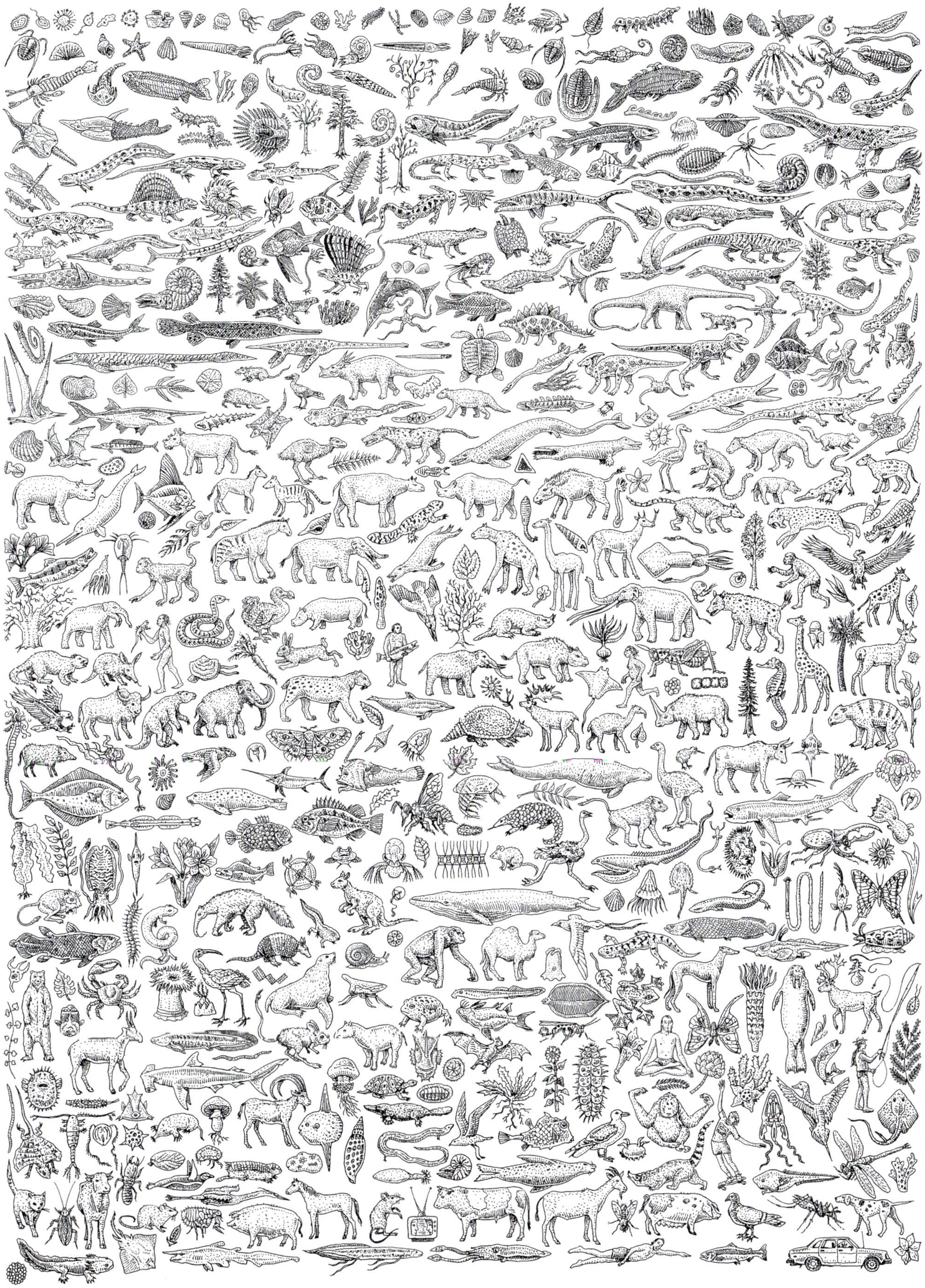

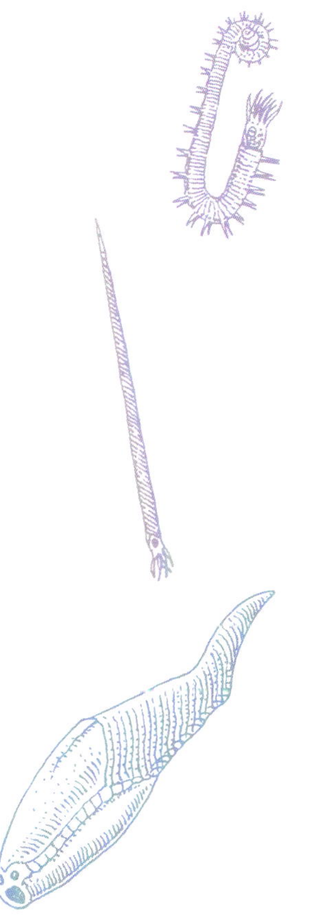

EACH LINE in the upper two-thirds of this drawing represents a geologic age and the animals and plants that flourished at that time. The bottom third represents modern-day life-forms—including, right along the bottom, a large number of domestic animals, insects, and yes, an automobile.

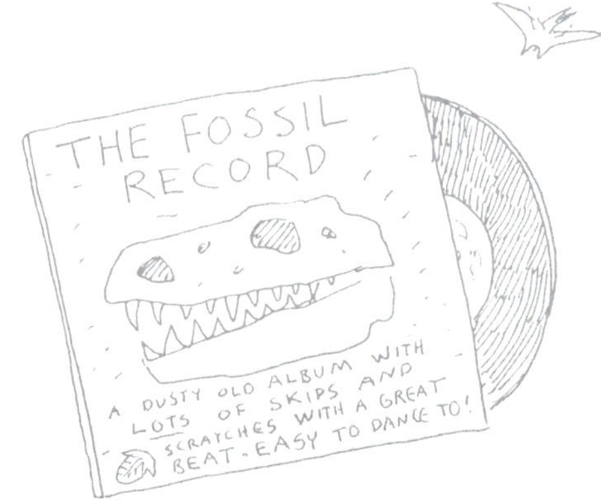

SCIENTISTS REFER TO the evidence of ancient life as the "fossil record." I stretched that phrase into a pun, ironic now that record albums are a sort of fossil themselves. The touring museum exhibit that grew out of the book *Planet Ocean* was called by this name. It featured a dance floor, an original soundtrack, amazing fossils, and all of the original art from the book.

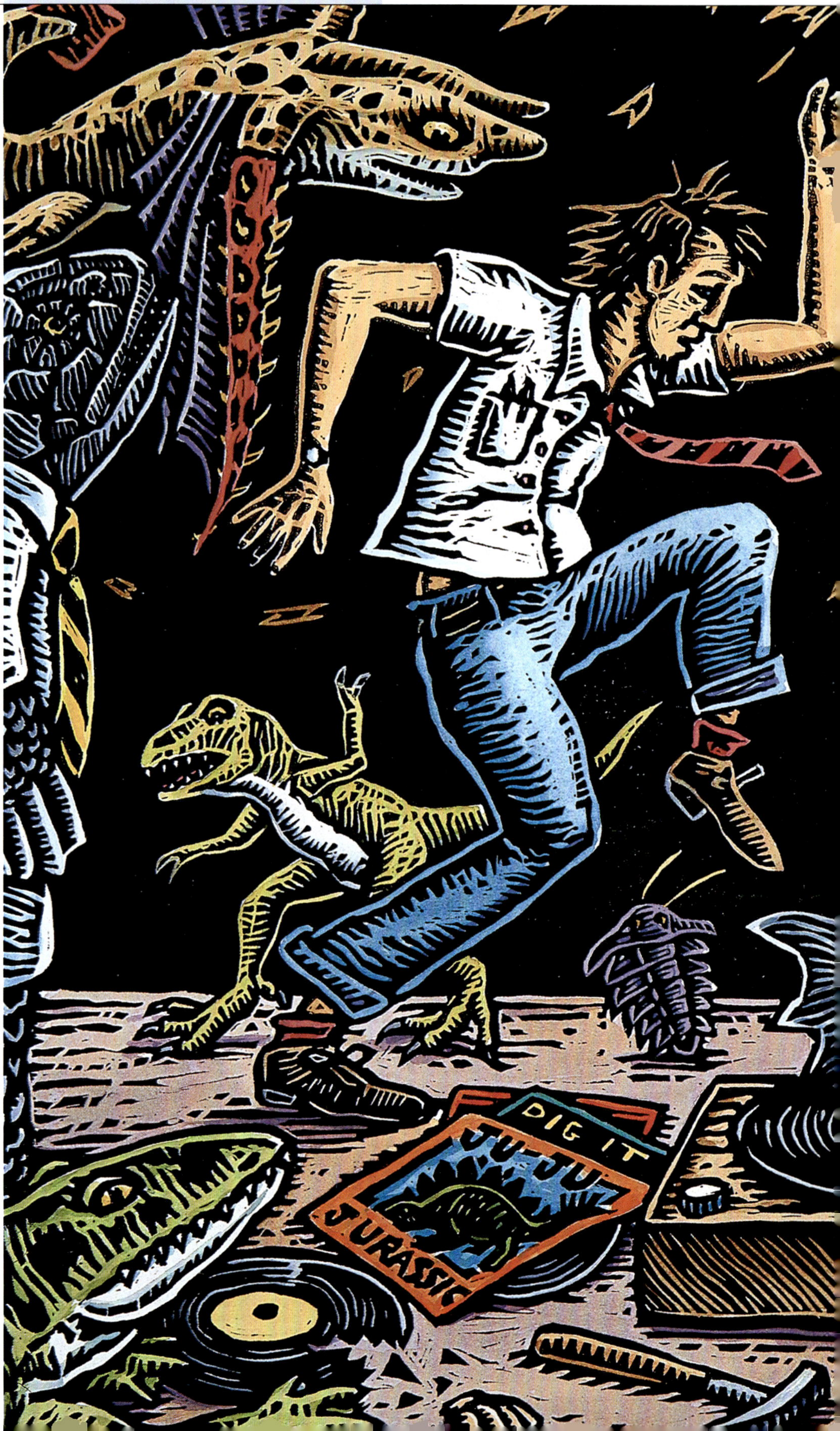

1995, linoleum block print and watercolor on paper, 18 in. × 24 in.

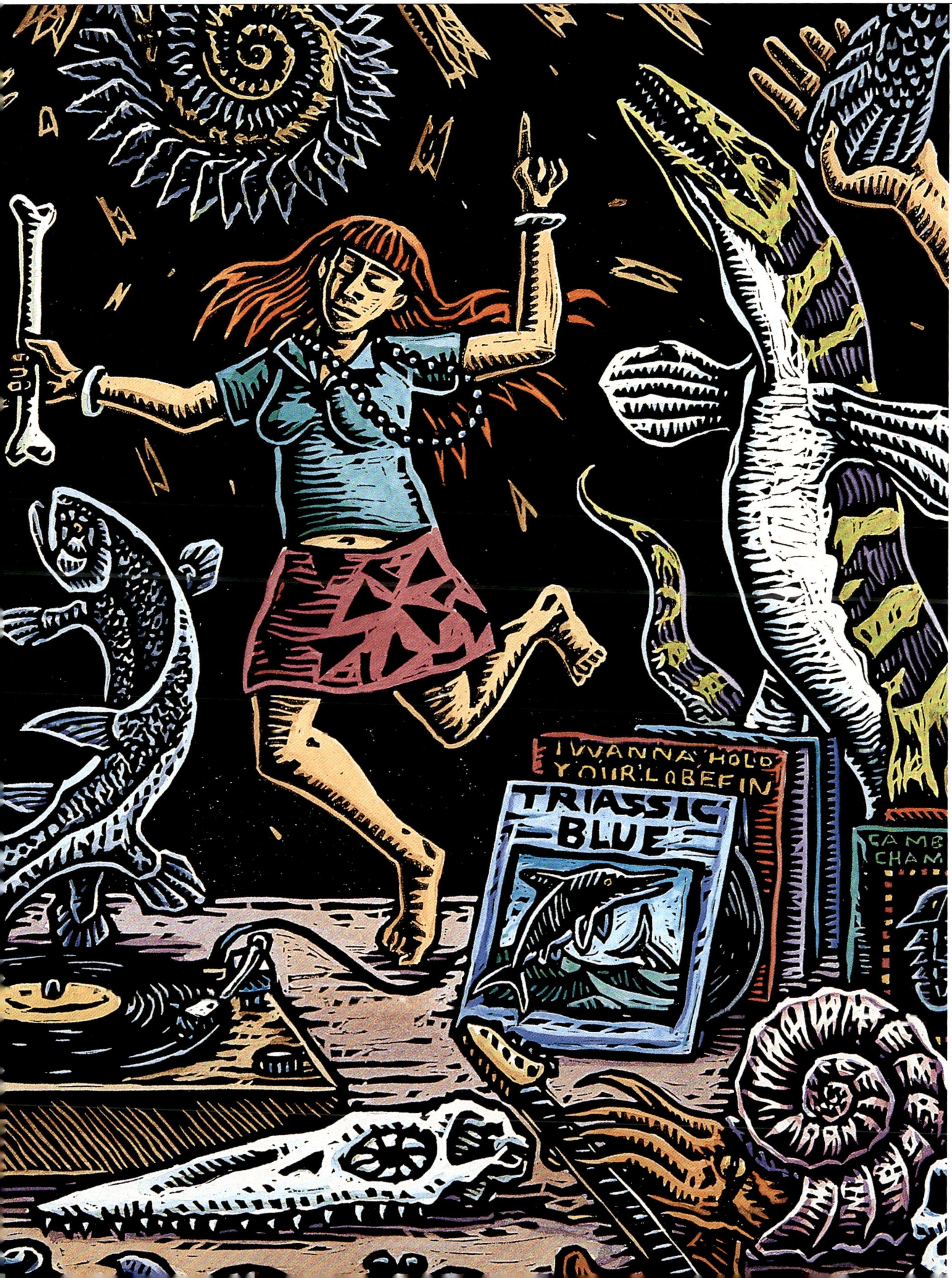

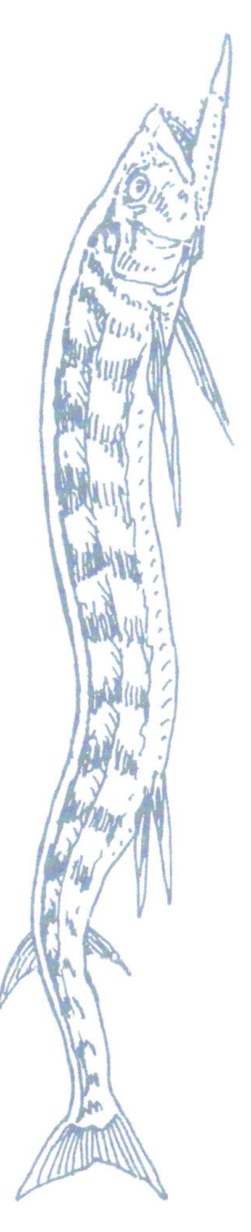

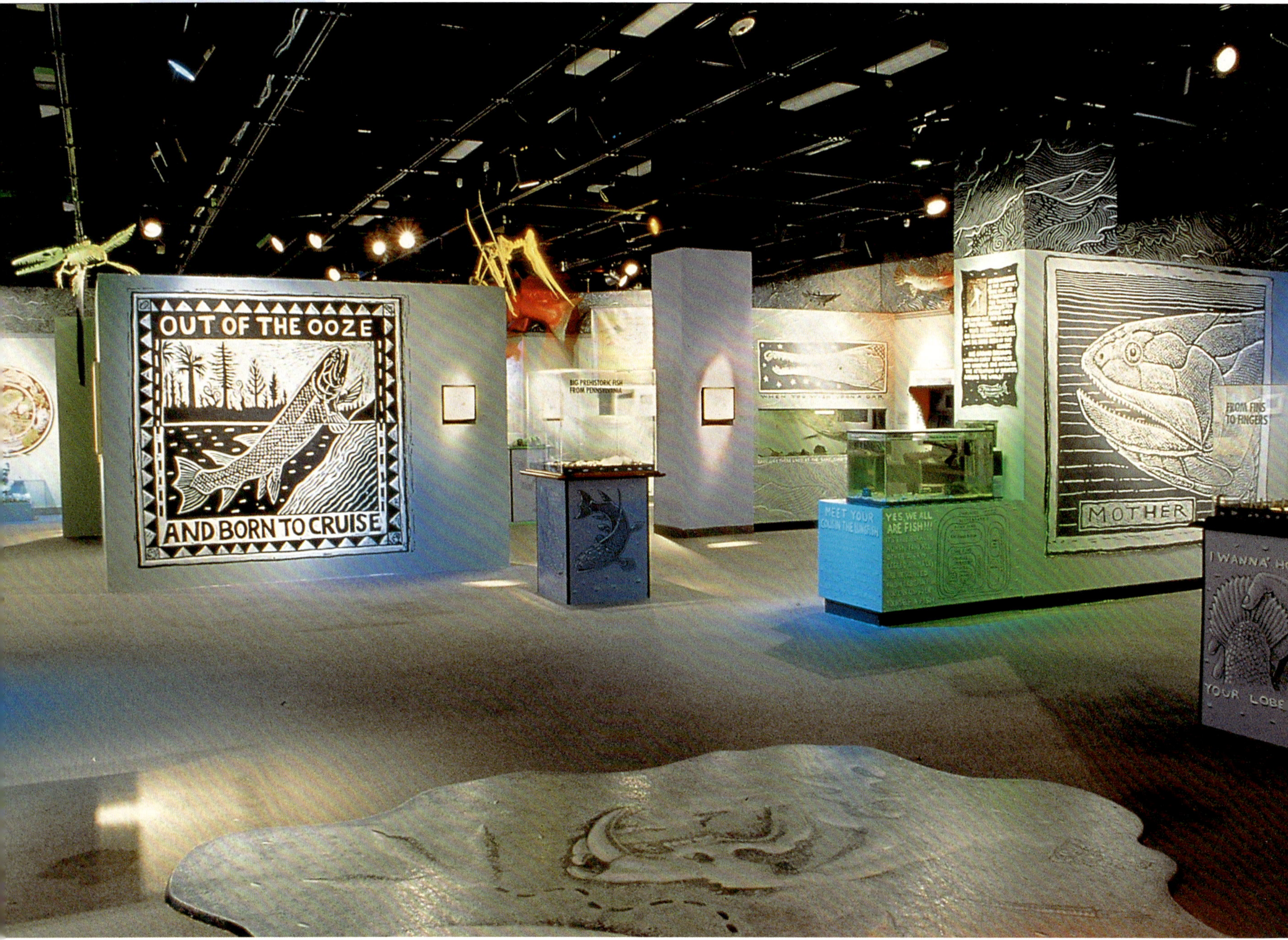

SEAN DURAN, in 1998 chair of traveling exhibits at the Academy of Natural Sciences, Philadelphia, and a crew of staff and volunteers helped make the Philly rendition of *Dancing to the Fossil Record* a totally immersive experience for museum visitors. We worked long hours painting virtually every square inch of space in the hall. The crowning touch was a gigantic organically shaped dance floor at the center of the room, painted in an improvisational style over one long weekend. I invited one of my all-time art heroes, Maryland artist Leonard Koscianski, to participate in that, and was blown away when he agreed to join in the fun. Shortly before we opened the doors for the exhibit, during the last few feverish hours of painting, one of my scientist heroes showed up to tour the hall—Stephen Jay Gould. Needless to say, I dropped my brush and showed him around.

Dancing to the Fossil Record traveled under a variety of names, with an ever-changing array of objects and installations. *Cruisin' the Fossil Freeway* was its new name at the Denver Museum of Nature and Science, 1999. "Night of the Giant Ammonites" was one of the new installations, complete with seventeen fossils arrayed below the mural.

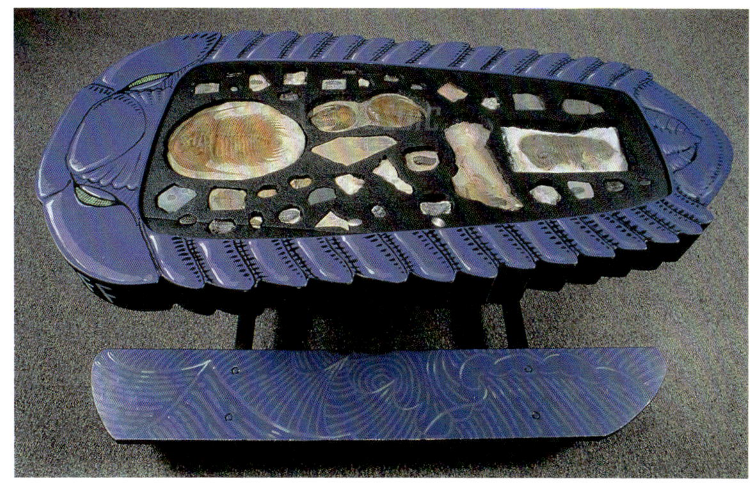

A stylish trilobite table at the Denver show.

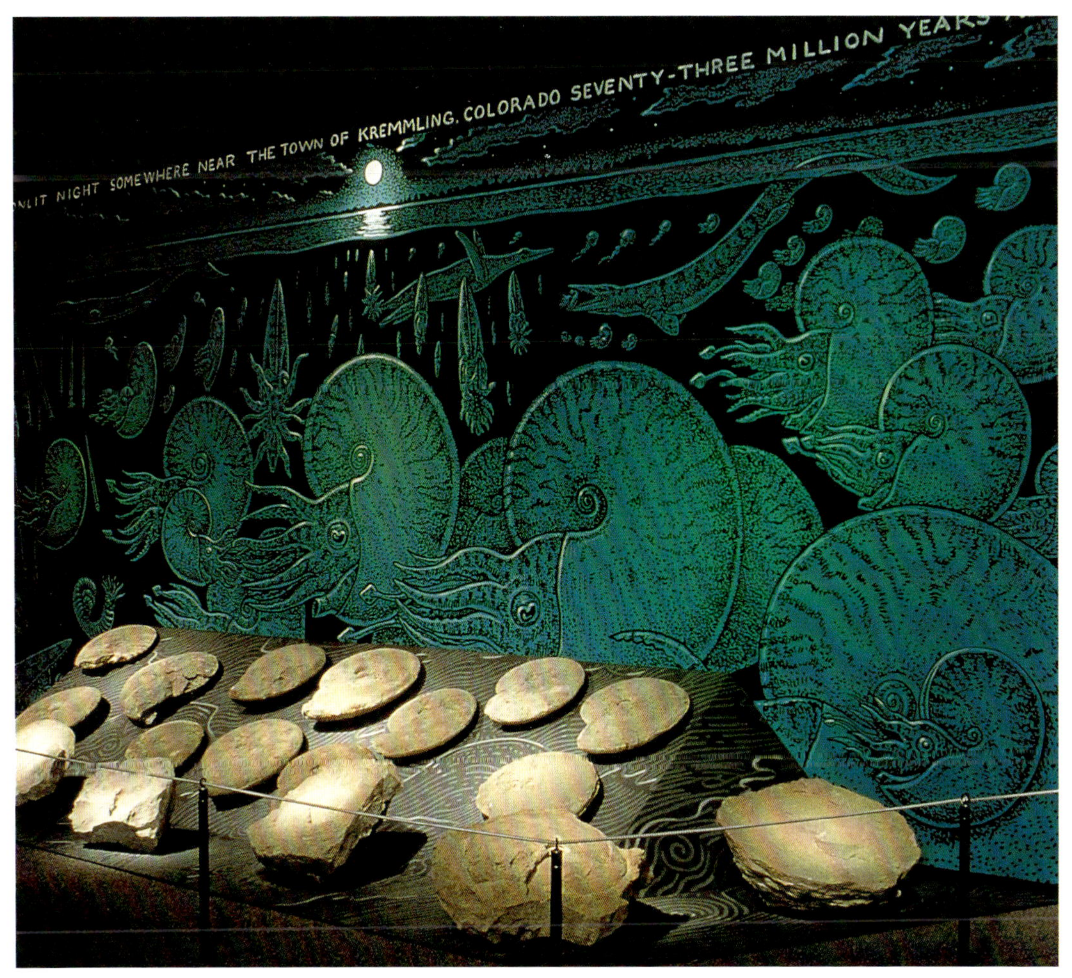

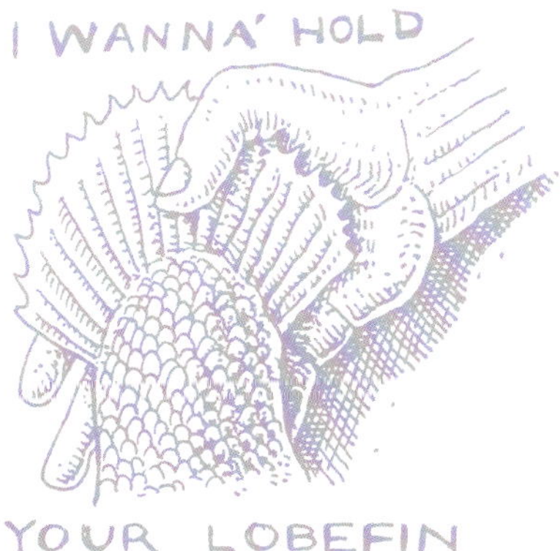

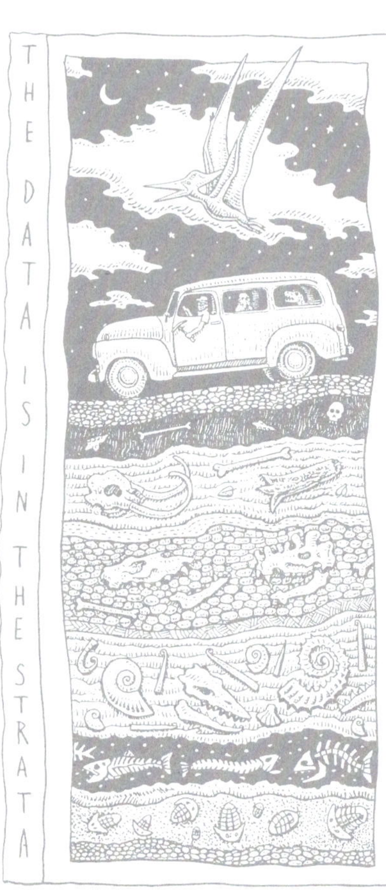

BAD OLD "Chuckie D" cruising in his stylin' Evolvo, spreading the good word about evolution.

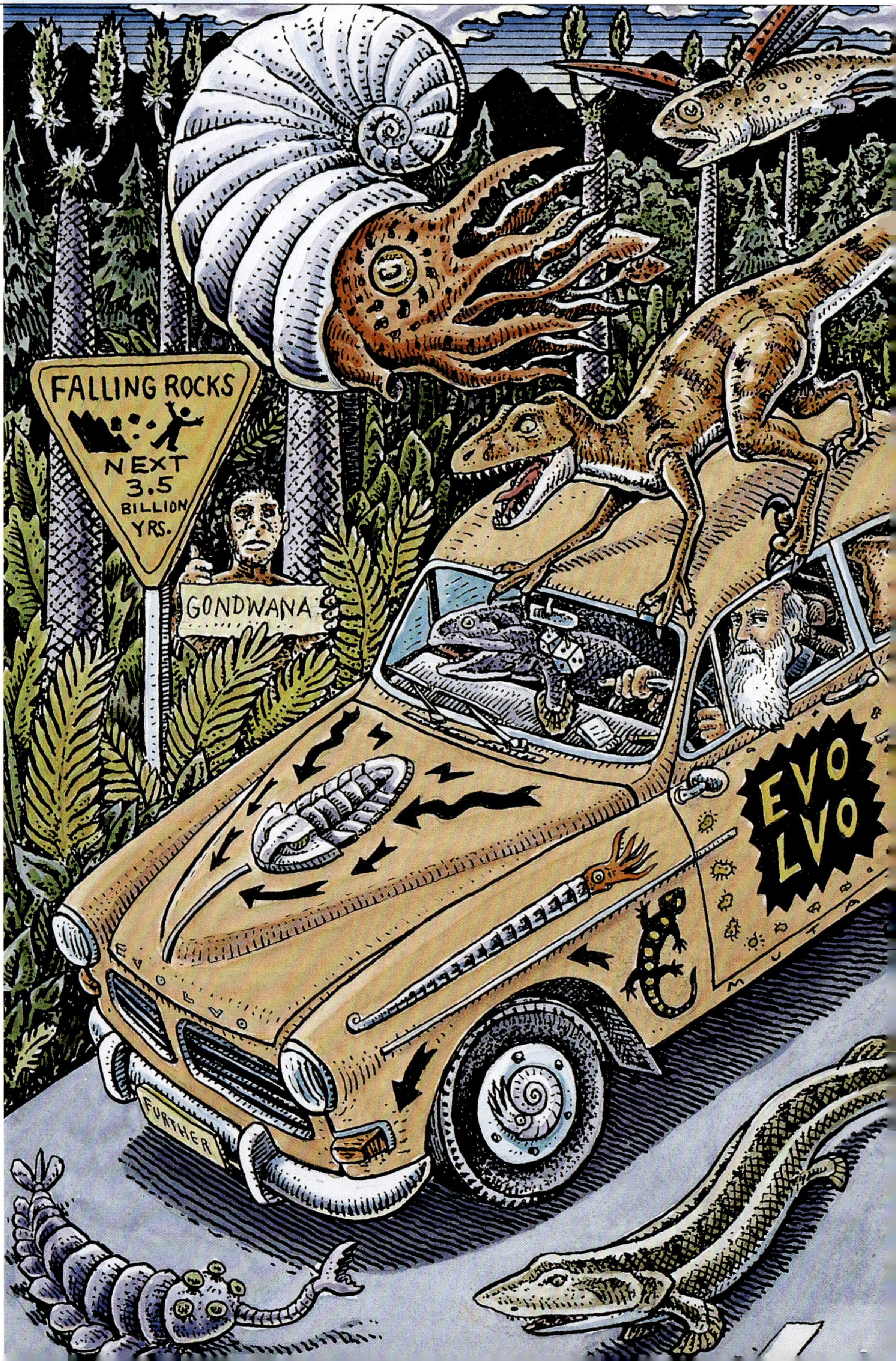

1998, pen, ink, and watercolor on paper, 8 in. × 8 in.

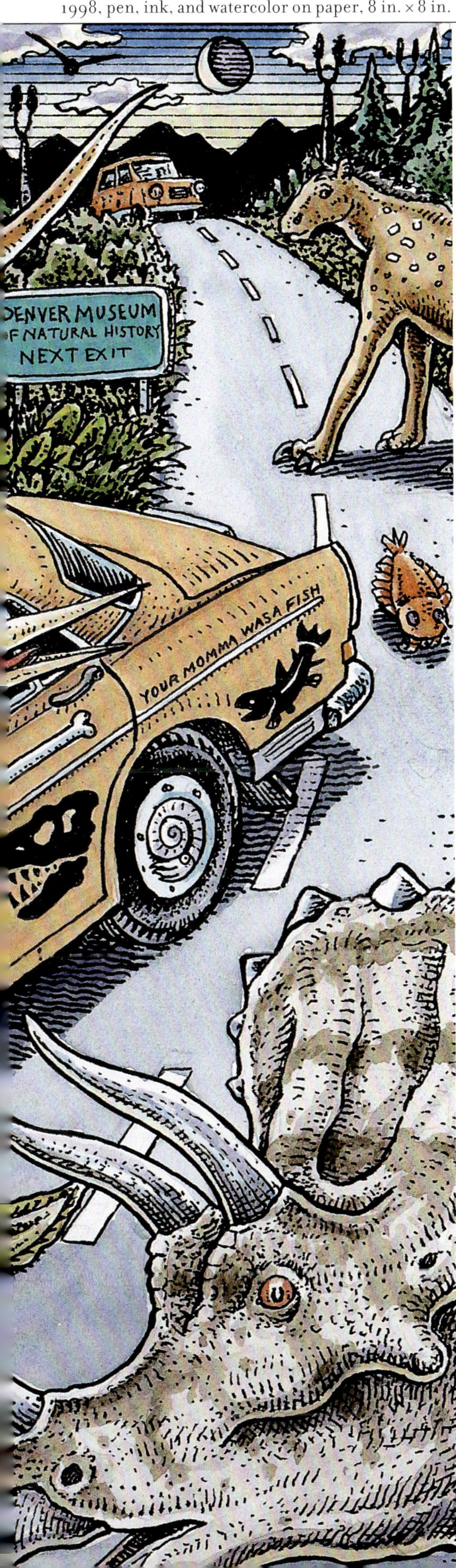

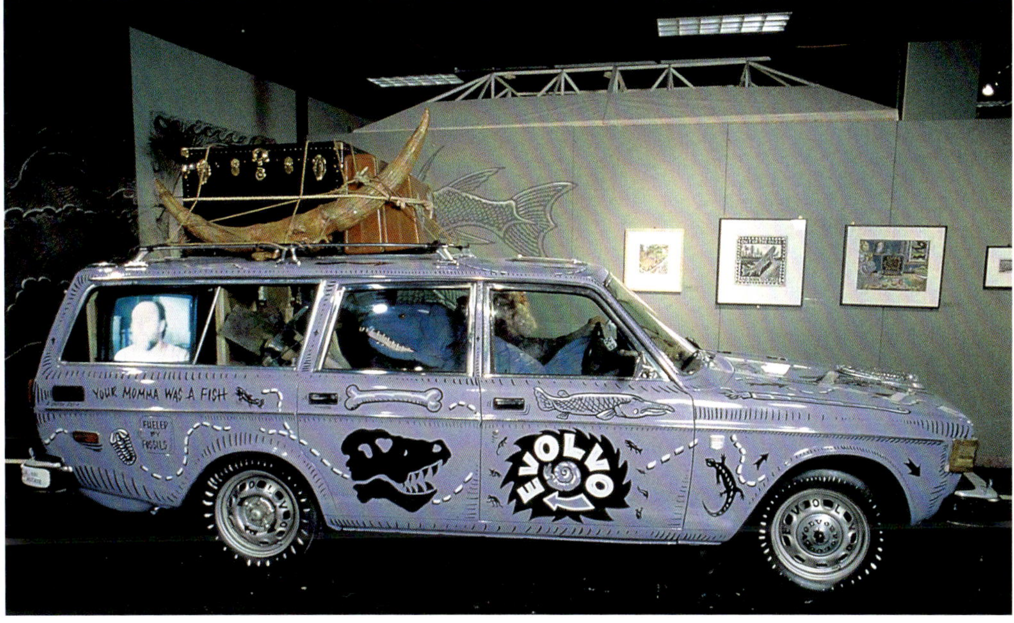

I PITCHED THE IDEA of an "Evolvo" art car to various institutions for many years; finally, in 1999, the staff at the Denver Museum of Nature and Science fulfilled my dream when they hosted my exhibit there. I acted as art director for Chuck Parson, a Denver-based artist, who tricked out the automobile with many of my icons. There was a talking-head video of me in the back explaining my work; in front, a lifelike sculpture of Charles Darwin sat at the wheel, while a lobe-finned fish *(Eusthenopteron)* occupied the passenger seat. The trilobite hood ornament was the sweetest touch. The current mayor of Denver, restaurateur/raconteur John Hickenlooper, is the proud owner of this art vehicle.

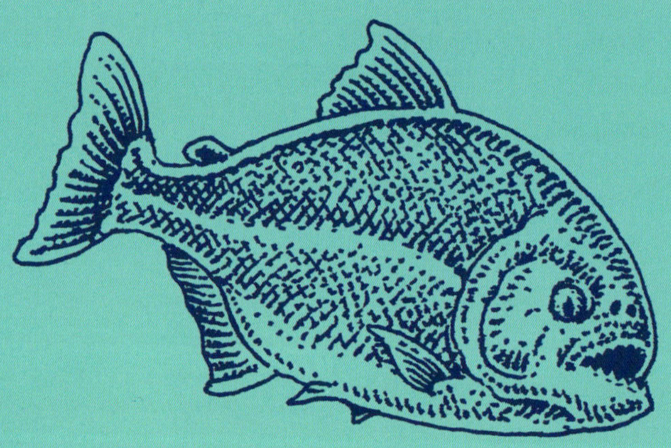

FISH STORM ON THE AMAZON

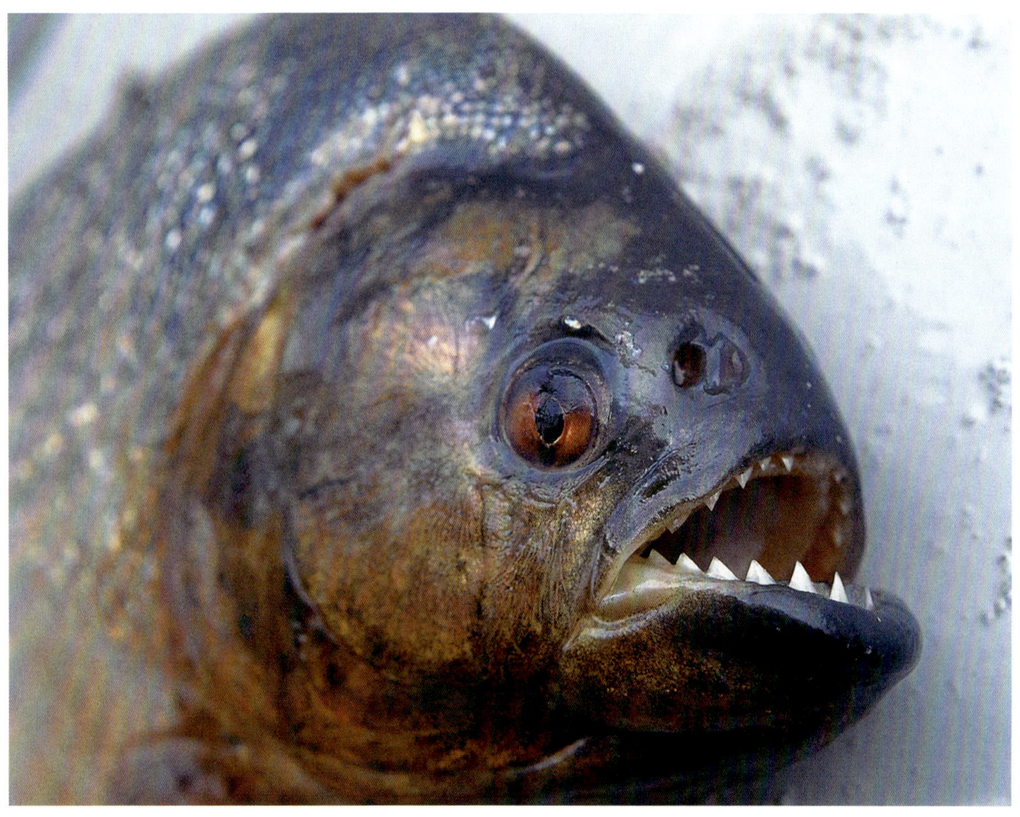

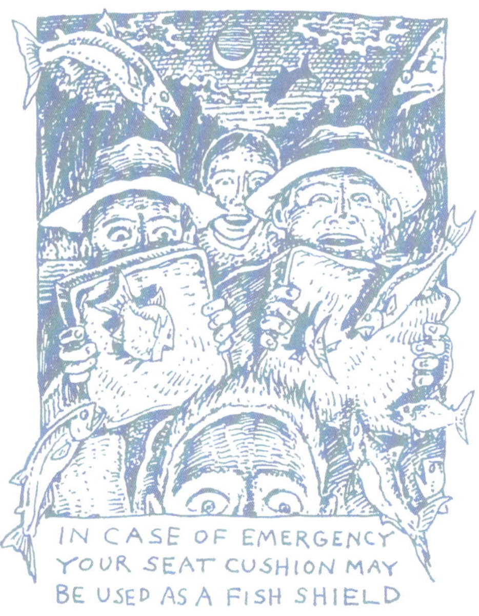

I WENT ON EXPEDITIONS to the Amazon River in 1997 and 2000 and have never been quite the same since. My paleo-mentor from the Denver Museum of Nature and Science, Kirk Johnson, has been organizing trips and filling boats with carefully selected adventure seekers for over a decade. Kirk told me the trips usually centered on fishing and he regaled me with tales for years. When he showed me a giant stack of exotic fish pictures, I knew I had to go. Kirk promised me it would be a fish-head's idea of nirvana, and he was absolutely right.

Above, we have a close-up of the notorious piranha, with razor-sharp teeth and blood-red eyes. There are several species of piranhas—this one is the black piranha, *Serrasalmus rhombeus*—and they are as ubiquitous in the Amazon as salmon are in Alaska. In fact, piranhas are *everywhere* in the Amazon, but they're rarely implicated in eating humans, and I've never found a valid report of anyone dying from a piranha attack. Although locals are sometimes bitten, it's usually in the process of removing one of these snapping devils from a fishing line. The toothy fish portrayed on the facing page is the infamous and undeservedly maligned red-bellied piranha.

2000, acrylic on board, 15 in. × 14 in.

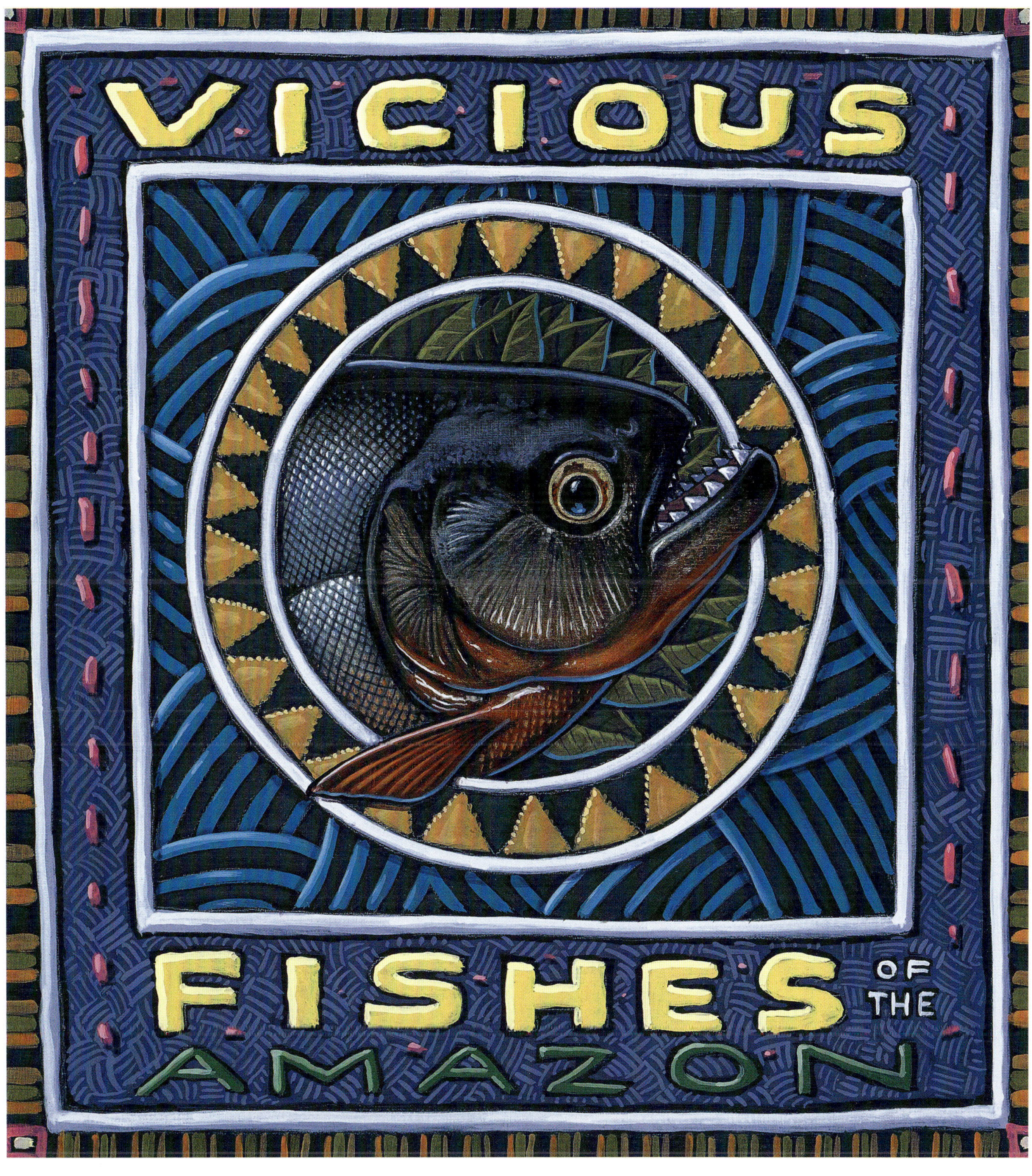

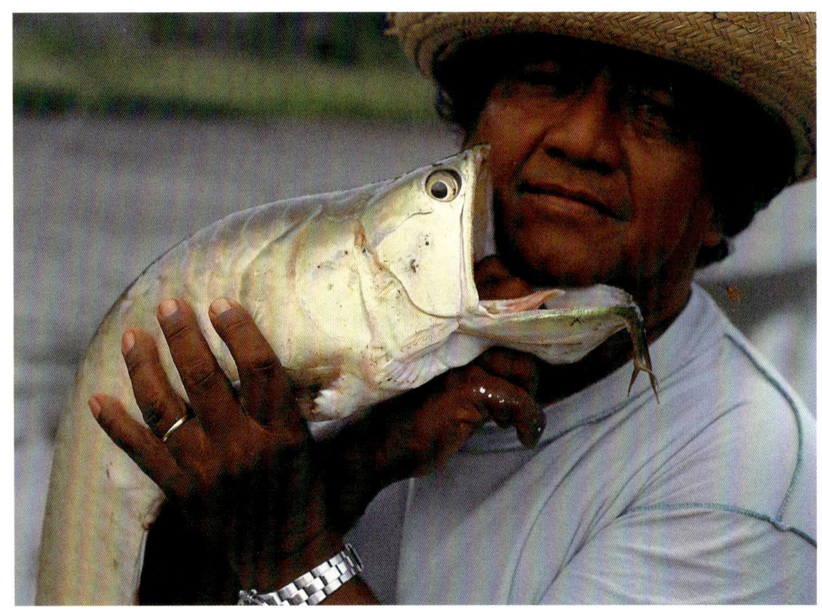

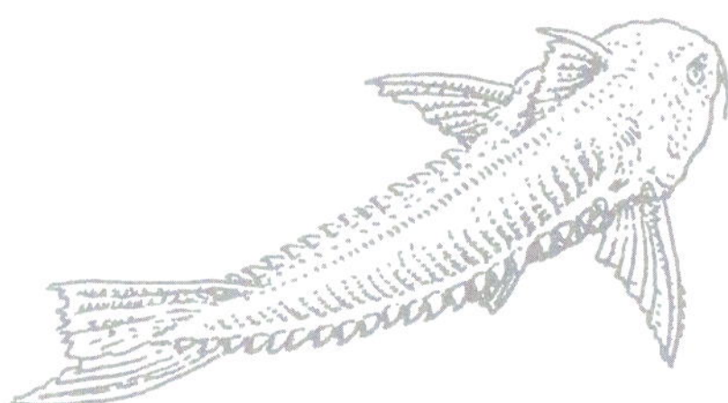

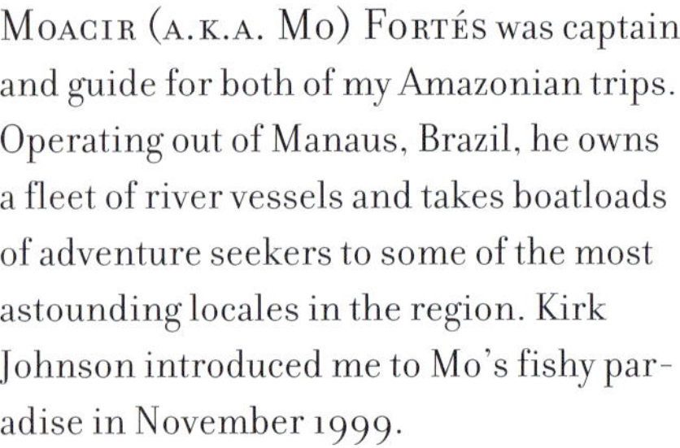

Moacir (a.k.a. Mo) Fortés was captain and guide for both of my Amazonian trips. Operating out of Manaus, Brazil, he owns a fleet of river vessels and takes boatloads of adventure seekers to some of the most astounding locales in the region. Kirk Johnson introduced me to Mo's fishy paradise in November 1999.

In this shot, Mo is holding up a freshly caught arowana. Earlier that day, we'd tried to catch one of these beautiful fish. They leapt all around us, and occasionally we'd see schools of them darting below our canoes, but although we cast into the shallows for hours, our efforts were in vain. We eventually pulled up alongside a fisherman's canoe and saw that he had an entire boatload of them (he had been using a gillnet). I asked Mo to hold one up by his face and snapped this portrait. The forked barbel hanging off the chin of the fish is used to feel along the surface of the water for insects, their favorite prey items.

We were cruising down the main channel of the Rio Solimoes, in a section of the river wider than an eight-lane superhighway, when we spotted some locals in a dugout canoe with this enormous catfish. Kirk and I hopped in Mo's canoe and pulled up alongside them to inspect the catch. This catfish, known locally as the piraíba, belongs to the genus *Brachyplatystoma.* It can grow to be up to six feet long—though local legend says it reaches ten feet or more and can swallow hapless bathers whole.

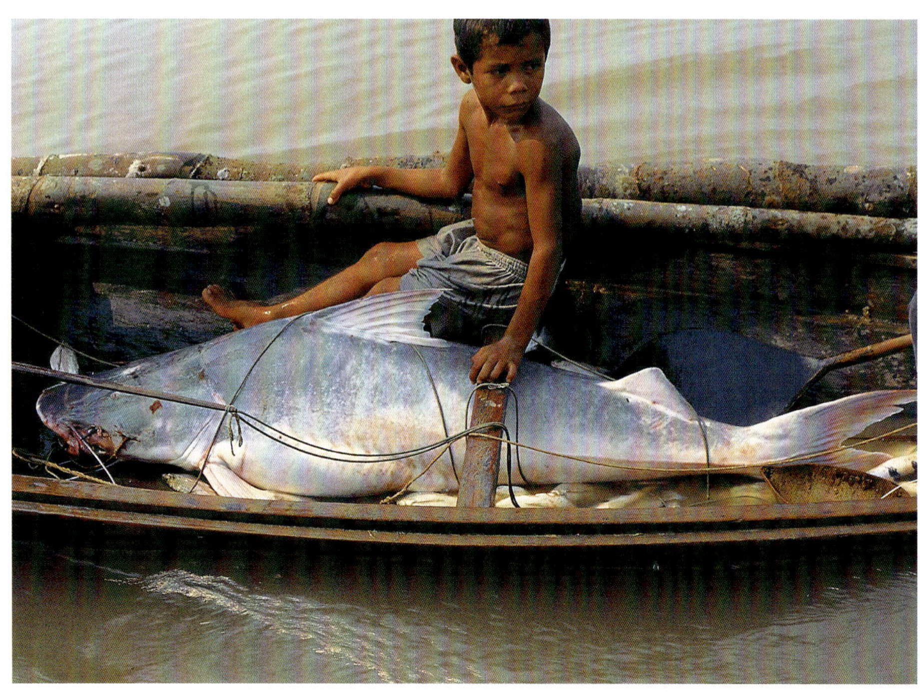

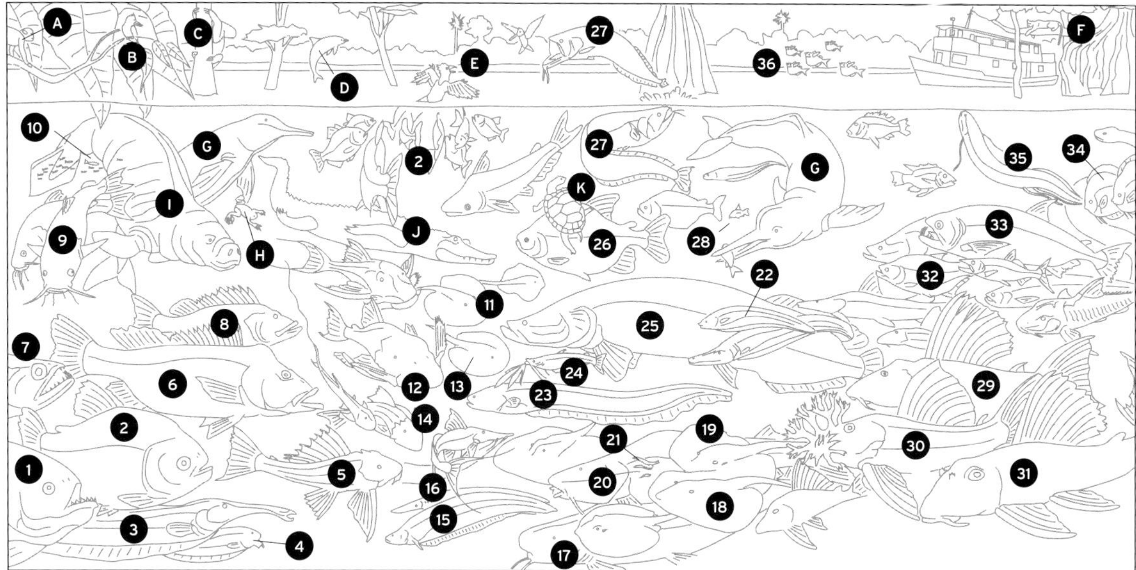

THE AMAZON is home to a vast array of creatures above and below the river's surface; the diversity of fishes alone rivals that of any ocean, with over three thousand known species—and scientists are finding more species every year. I painted this seven-by-fifteen-foot-long mural over the course of nine months in 1999–2000, filling it with pink dolphins, a manatee, a curious squirrel monkey, a large black caiman, an anaconda, and a paradisiacal host of fishes. Brad Matsen and I can be seen on the back deck of the boat. I have a fishing rod in my hand and am wearing a life jacket, revealing my cowardly tendency to panic in dangerous situations. Brad is standing fearlessly with an Antarctica beer in his hand.

I've been working for the last few years with my good friend and museum exhibits guru Sean Duran of the Miami Museum of Science to develop an NSF-funded traveling show called *Amazon Voyage: Vicious Fishes and Other Riches;* this mural will be a centerpiece of the show. It has already gone out into the world as my first national magazine cover ever, gracing a 2001 issue of *Natural History.* Dr. John Lundberg, a noted authority on South American fishes, wrote the accompanying article.

FISH . . .

1. Black piranha, *Serrasalmus rhombeus*
2. Red-bellied piranha, *Pygocentrus nattereri*
3. Thin-nosed tube-snout knifefish, *Sternarchorhynchus oxyrhynchus*
4. *Sternarchogiton* sp.
5. Dark-spotted armored catfish, *Pterygoplichthys* sp.
6. Speckled peacock bass, *Cichla monoculus*
7. Freshwater dogfish, *Rhaphiodon vulpinus*
8. Peacock bass, *Cichla temensis*
9. Pirarara (red-tailed) catfish, *Phractocephalus hemioliopterus*
10. Cardinal tetra, *Paracheirodon axelrodi*
11. Motoro, *Potamotrygon* sp.
12. Royal pleco, *Panaque nigrolineatus*
13. Reticulated stingray, *Potamotrygon reticulatus*
14. Common pleco, *Hypostomus* sp.
15. Duke's channel knifefish, *Magosternarchus duccis*
16. Tamandua tube-snout knifefish, *Orthosternarchus tamandua*
17. Black piraiba catfish, *Brachyplatystoma* sp.
18. Tiger stingray, *Potamotrygon* sp.
19. Faulkner's stingray, *Potamotrygon faulkneri*
20. Piraiba catfish, *Brachyplatystoma filamentosum*
21. Candirú catfish, *Vandellia plazaii*
22. Tooth-lip knifefish, *Oedemognathus exodon*
23. Electric eel, *Electrophorus electricus*
24. *Hypophthalmus fimbriatus*
25. Pirarucú, *Arapaima gigas*
26. Tambaqui, *Colossoma macroponum*
27. Arowana, *Osteoglossum bicirrhosum*
28. Harald Schultz's cory, *Corydoras haraldschultzi*
29. *Hypostomus* sp.
30. Polka-dot bristlenose catfish, *Ancistrus lineolatus*
31. Goldy pleco, *Scobiancistrus aureatus*
32. Elongate piranha, *Serrasalmus elongatus*
33. Pirandirá vampirefish, *Hydrolycus scomberoides*
34. Heckel discus, *Symphysodon discus*
35. American lungfish, *Lepidosiren paradoxa*
36. Hatchetfish, *Thoracocharax stellatus*

. . . AND FRIENDS

A. Squirrel monkey, *Saimiri sciureus*
B. Green kingfisher, *Chloroceryle americana*
C. Three-toed sloth, *Bradypus tridactylus*
D. Tucuxi dolphin, *Sotalia fluviatilis*
E. Hoatzin, *Opisthocomus hoazin*
F. Jaguar, *Panthera onca*
G. Amazon river dolphin, *Inia geoffrensis*
H. Surinam toad, *Pipa pipa*
I. Amazonian manatee, *Trichechus inunguis*
J. Spectacled caiman, *Caiman crocodilus*
K. Yellow-spotted Amazon turtle, *Podocnemis unifilis*

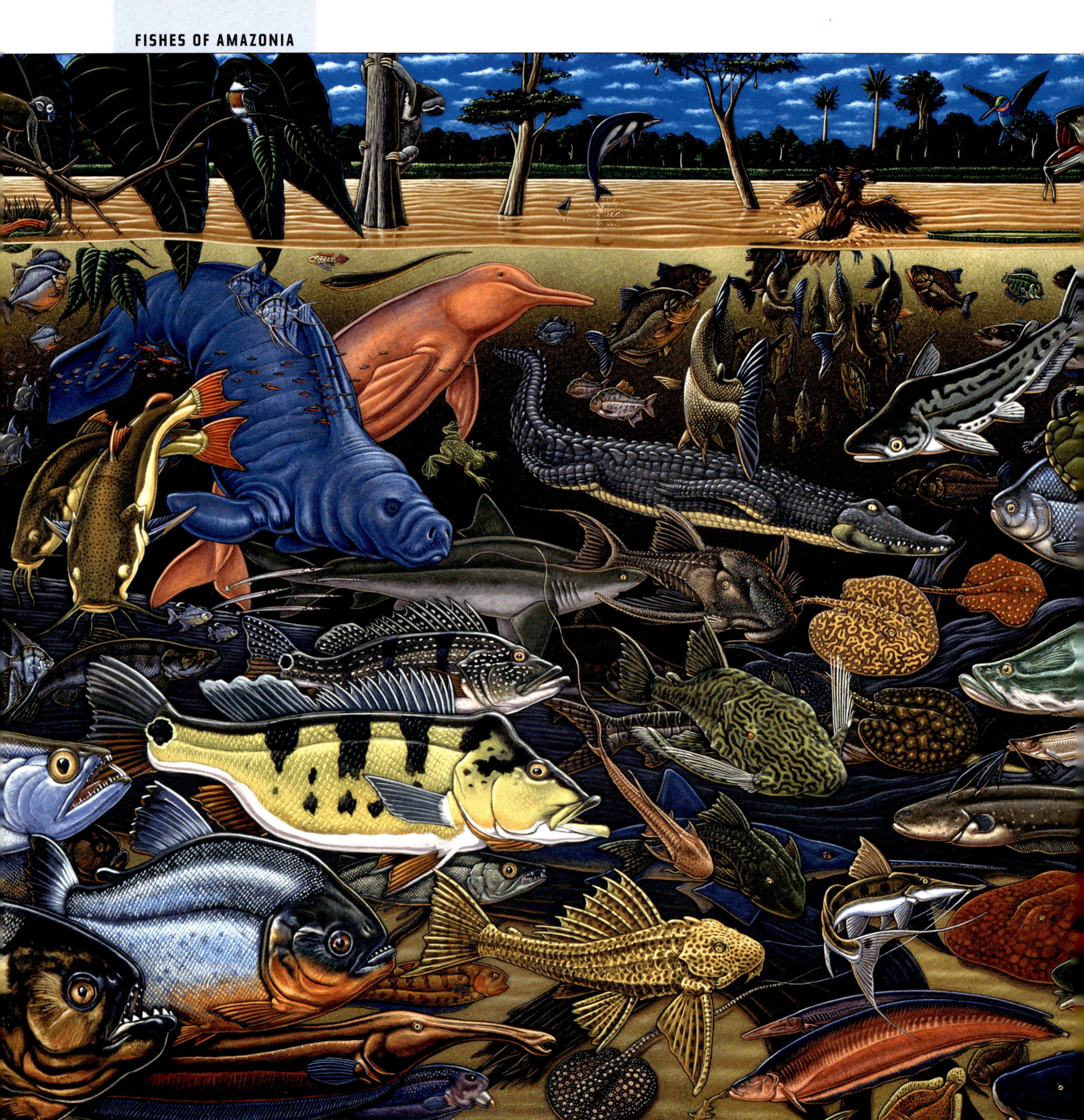

2000, acrylic on canvas, 84 in. × 180 in.

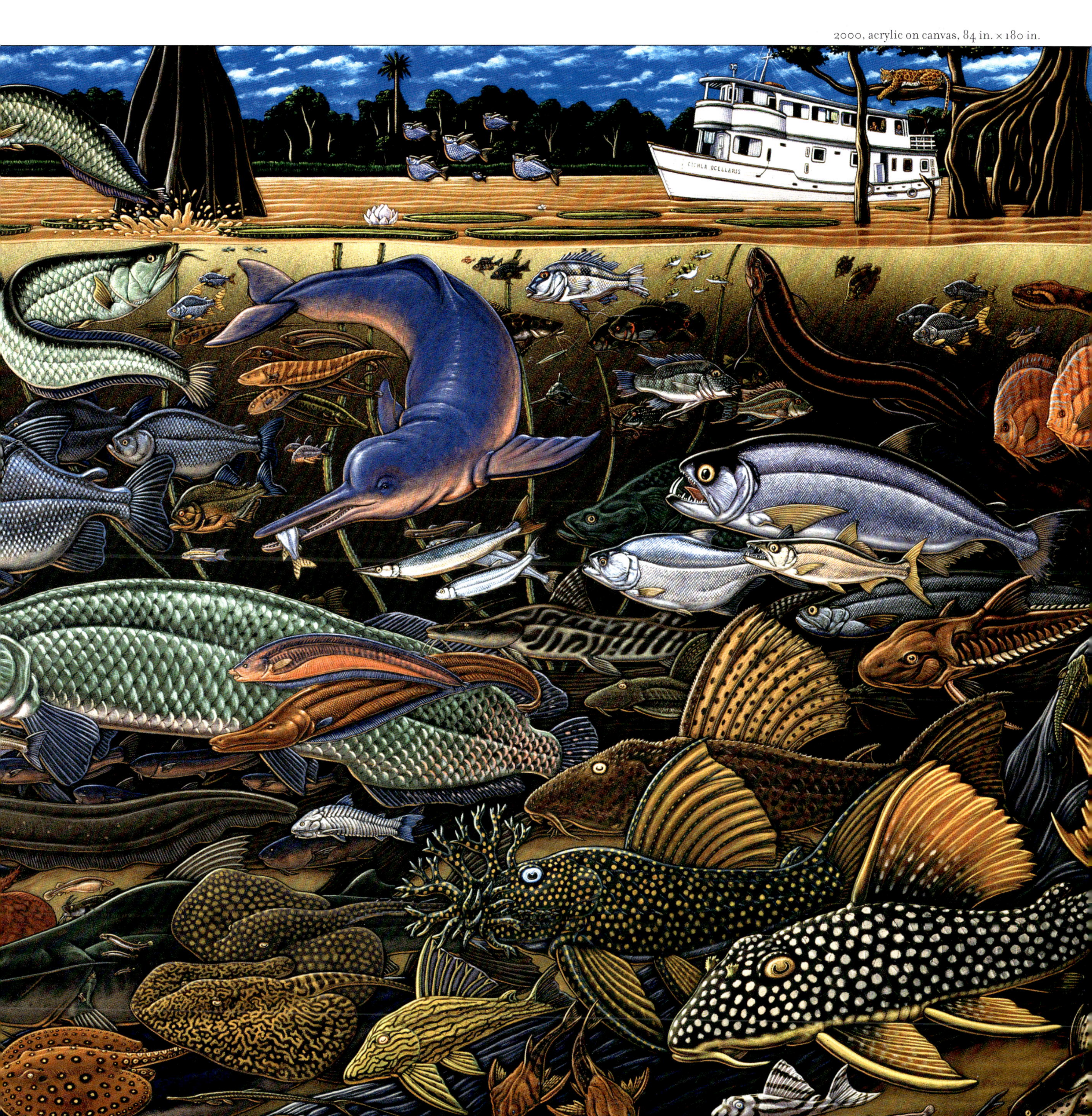

RIVER RAYS

2001, colored pencil and acrylic paint on paper, 22 in × 55 in.

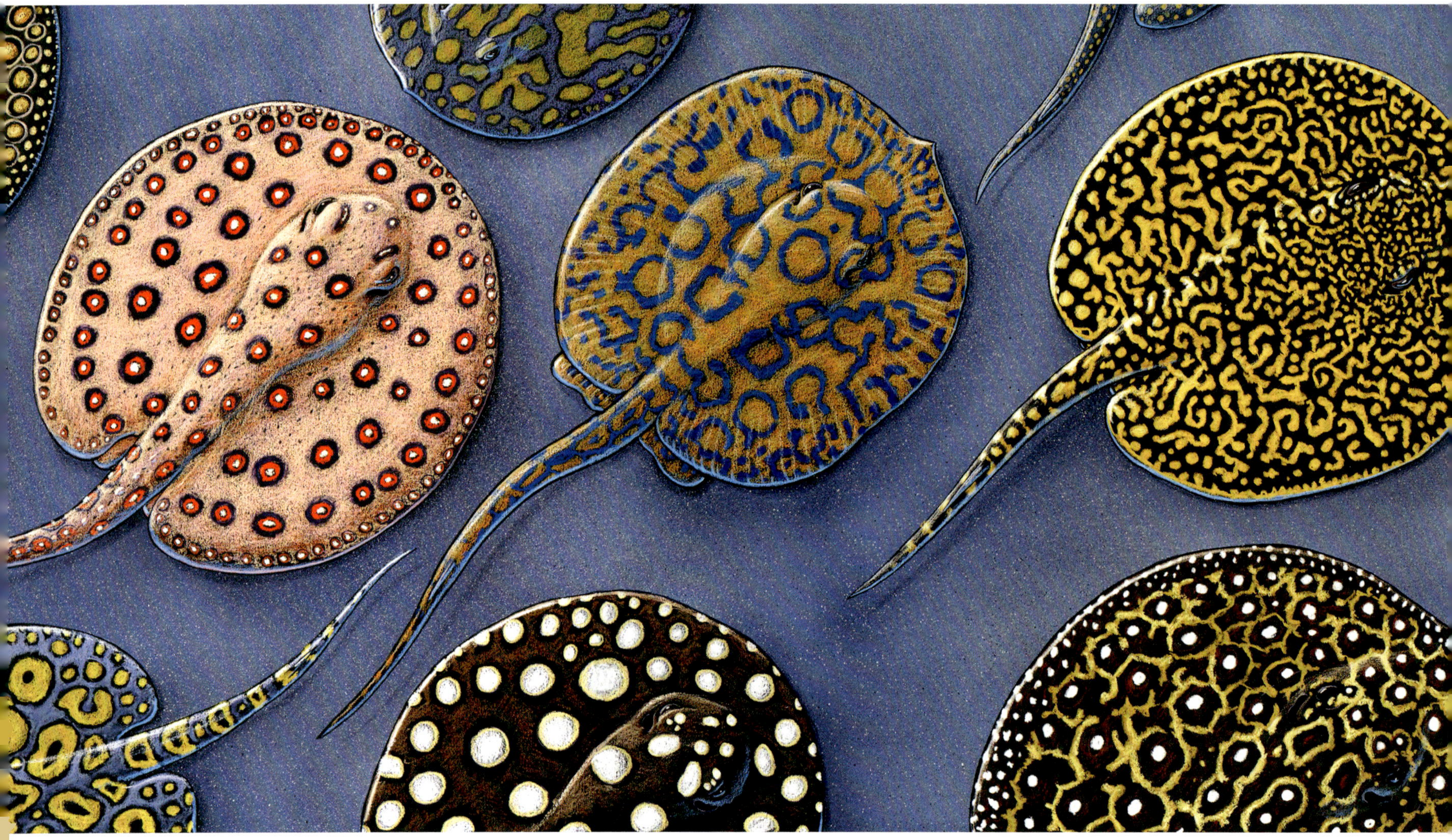

STINGRAYS ARE FOUND all over the Amazon River basin. There are over twenty different species, most of which are adorned with spectacular camouflage patterns. Thanks to their poisonous stingers, which pack a powerful punch, they are feared by the locals more than piranhas. When I was in Brazil I quickly learned to do the "stingray shuffle" when wading in shallow water, kicking the sand in front of me before plunking my foot down.

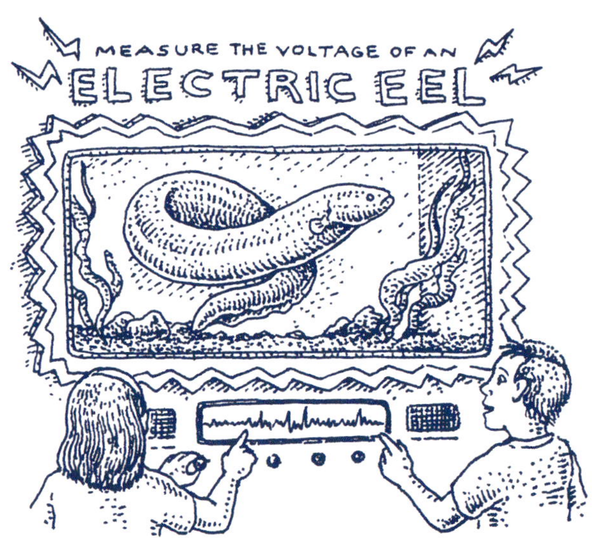

REMEMBERING GONDWANA

2000, acrylic on board, 20 in. × 9 in.

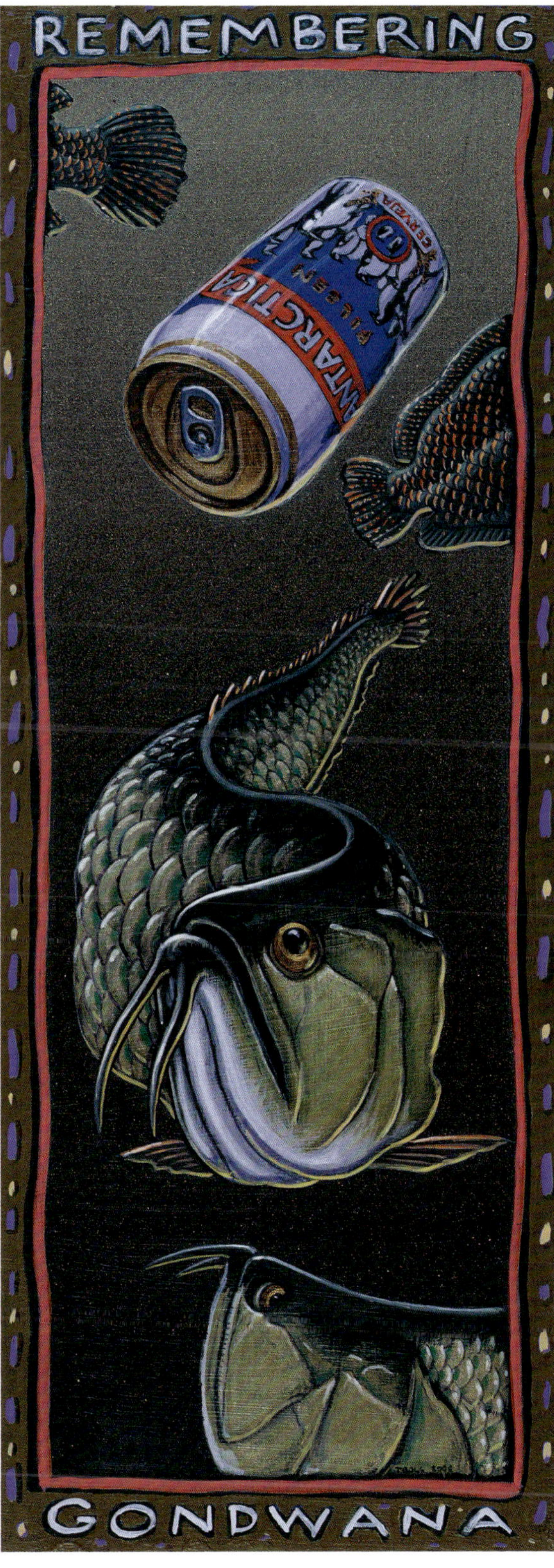

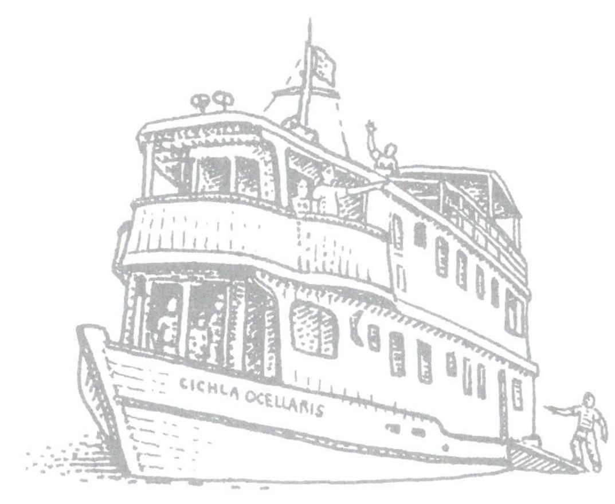

WHILE TRAVELING by riverboat in Brazil, I thought it surreal that the local beer of choice was named Antarctica. The running joke on our scientist-filled boat was that it was an oblique tribute to the ancient southern supercontinent called Gondwana. One hundred thirty million years ago, South America, Antarctica, Australia, Africa, and India were all part of one gigantic landmass. One of the clues to solving the geologic puzzle of moving continents came from cataloging similar freshwater fishes across different, now-separated continents.

The fish contemplating the beer can in this painting is an arowana. Found throughout the Amazon basin, locally it's called a "monkey fish" because it makes spectacular leaps to snag unwary prey sitting in branches above the river. Related species are also found in Africa and Australia. Since there's no way a freshwater fish could have migrated between these widely separated continents, the only explanation could be that the continents were once joined. It's also an indication that the arowana lineage is very, very old.

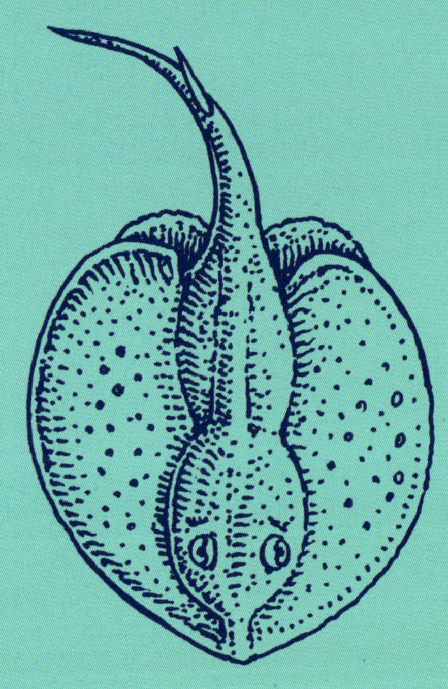

SHARKS AND THEIR RELATIVES

1986, colored pencil on paper, 17 in. × 22 in.

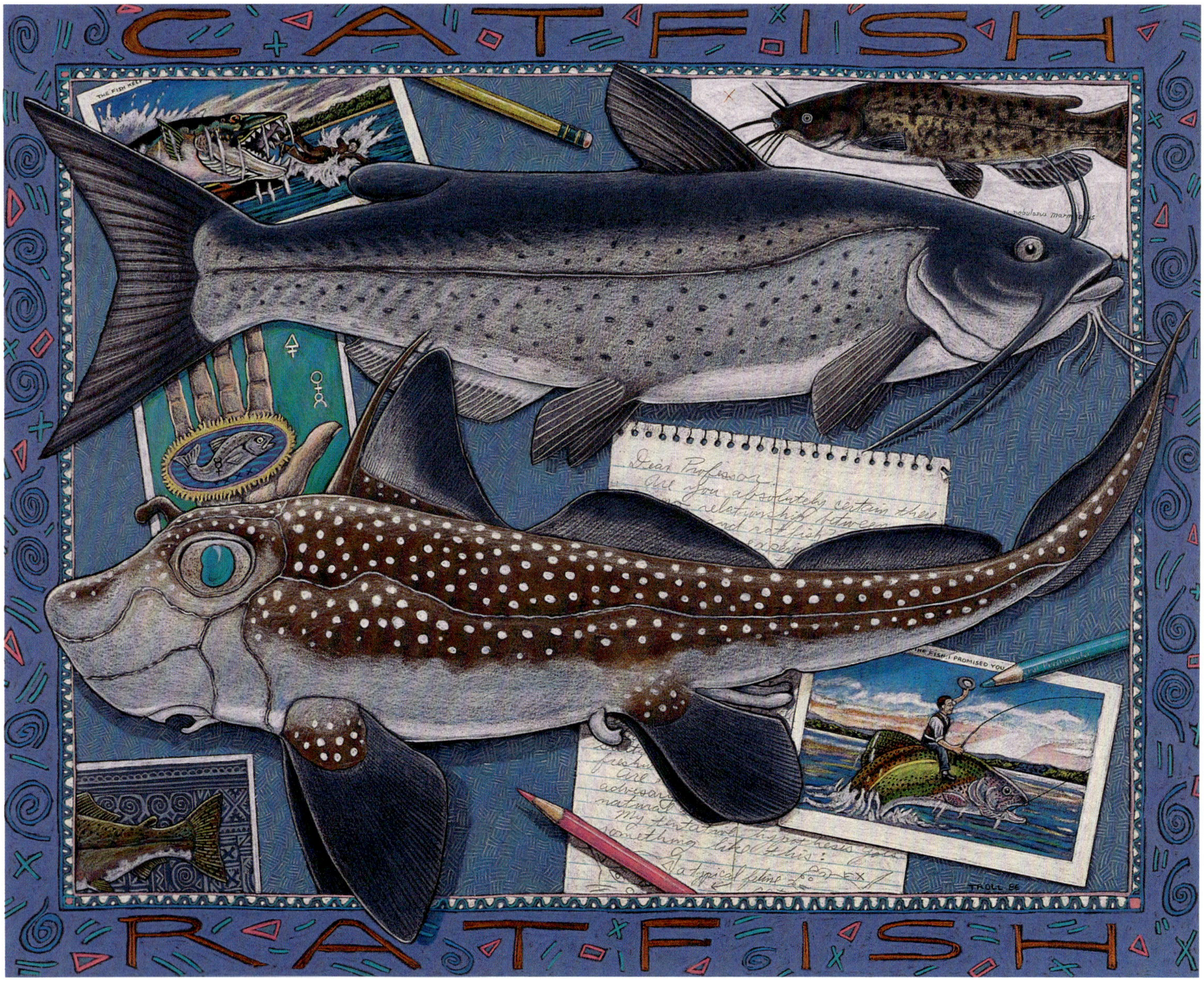

A BONY FISH at the top, a cartilaginous fish at the bottom. This is an early drawing that coincidentally foreshadows two passions that emerged later in my work: the Amazon and sharks.

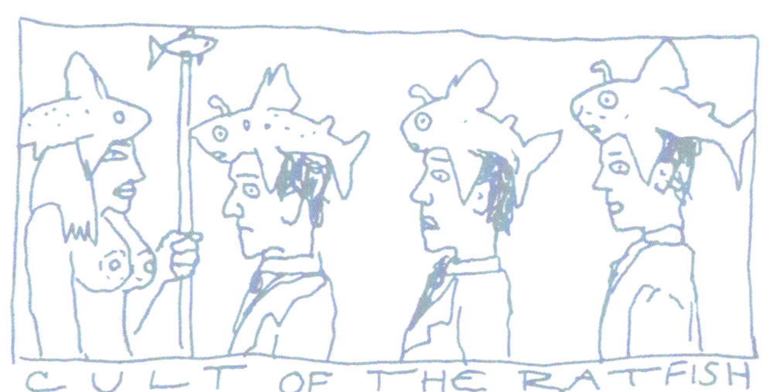

2001, colored pencil on paper, 16 in. × 24 in.

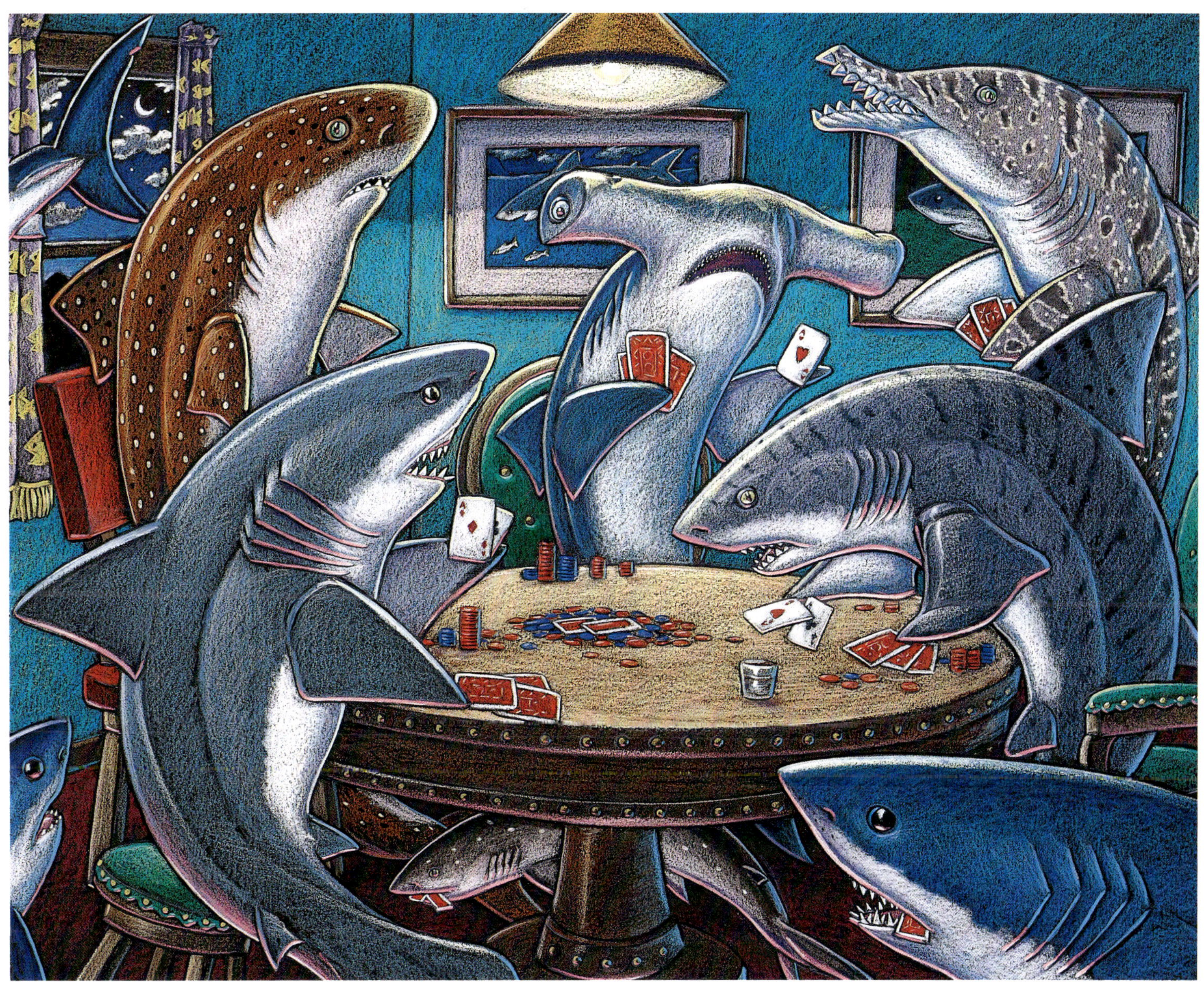

THE AMERICAN lowbrow masterpiece *Dogs Playing Poker* has always exerted a forbidden appeal on my trashy art sensibility. I indulged that itch for kitsch when I did this not-so-subtle homage.

When I first caught a ratfish about twenty-odd years ago, I knew there was something special about this fish. With its turquoise eyes, large head, and translucent fins, it looked unlike anything else in the ocean I had seen before. The males have bizarre appendages on their heads called tenaculums, which they use to grab the girls in the "dance of life." Well-defined sensory canals cover their bodies, making them look like a science experiment gone wrong. They're also known as chimeras, an appropriate moniker as it's the name of a monster in Greek mythology made from the bodies of different animals. I was completely transfixed by this otherworldly fish. Besides, there is no hipper name for a fish than "ratfish."

My interest soon turned into an unhealthy obsession as I began to correspond with ratfish specialists around the globe (an admittedly limited group of individuals).

The path led me to Dr. Dominique Didier Dagit in Philadelphia, arguably the foremost authority on chimeras on the planet. I call her D3-RFQ (ratfish queen!).

D3-RFQ and I corresponded for years, talked often on the phone, and eventually met to plot a global ratfish consciousness-raising campaign. My appreciation for ratfish deepened when I understood that they are *truly* ancient creatures, dating far back in geologic time and virtually unchanged for 325 million years. These "living fossils" were cruising the depths of the oceans long before dinosaurs roamed the land.

Chimeras are cartilaginous fish, having a skeleton made not of bone but of cartilage. This characteristic, among others, puts them firmly in the shark family tree. They're the sole surviving members of an ancient lineage that was once a very diverse branch of that tree. I like to think of them as living Paleozoic sharks.

D3-RFQ honored my ratfish awareness efforts a few years ago by naming a newly discovered ratfish species from New Zealand after me: *Hydrolagus trolli*—meaning, literally, "water rabbit troll." This is a drawing of that species. The two fish at the bottom are males.

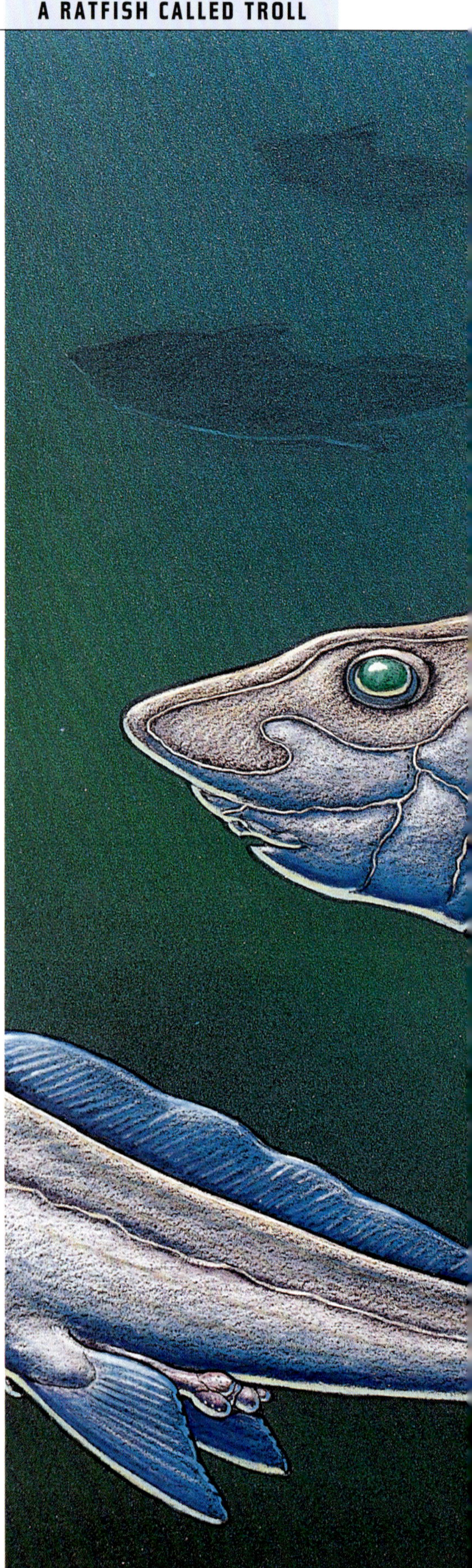

2002, colored pencil and acrylic paint on paper, 22 in. × 30 in.

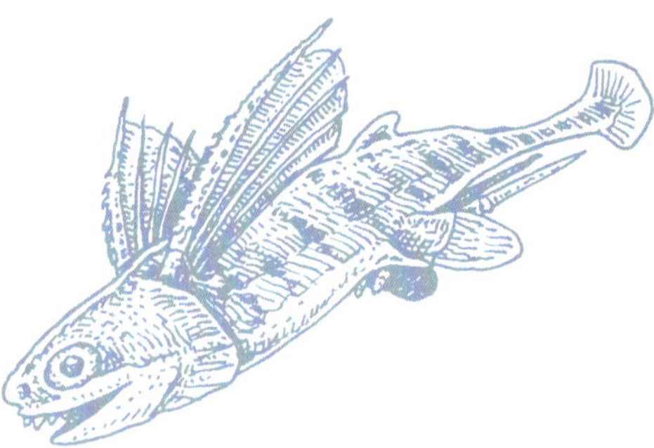

In the Paleozoic era sharks and their kin were an incredibly diverse group of animals, with all kinds of strange-looking species abounding. These little weirdos are members of an extinct order of shark relatives called the Iniopterygians, which means "neck fin." The odd positioning of the large pectoral fins high on their bodies has led some scientists to speculate that they could glide through the air like modern-day flying fish.

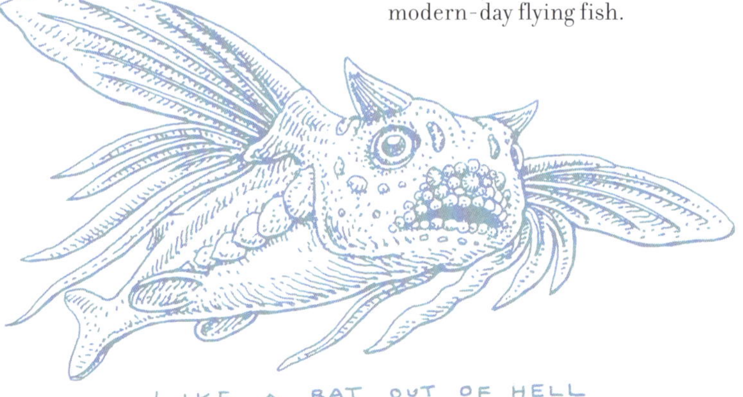

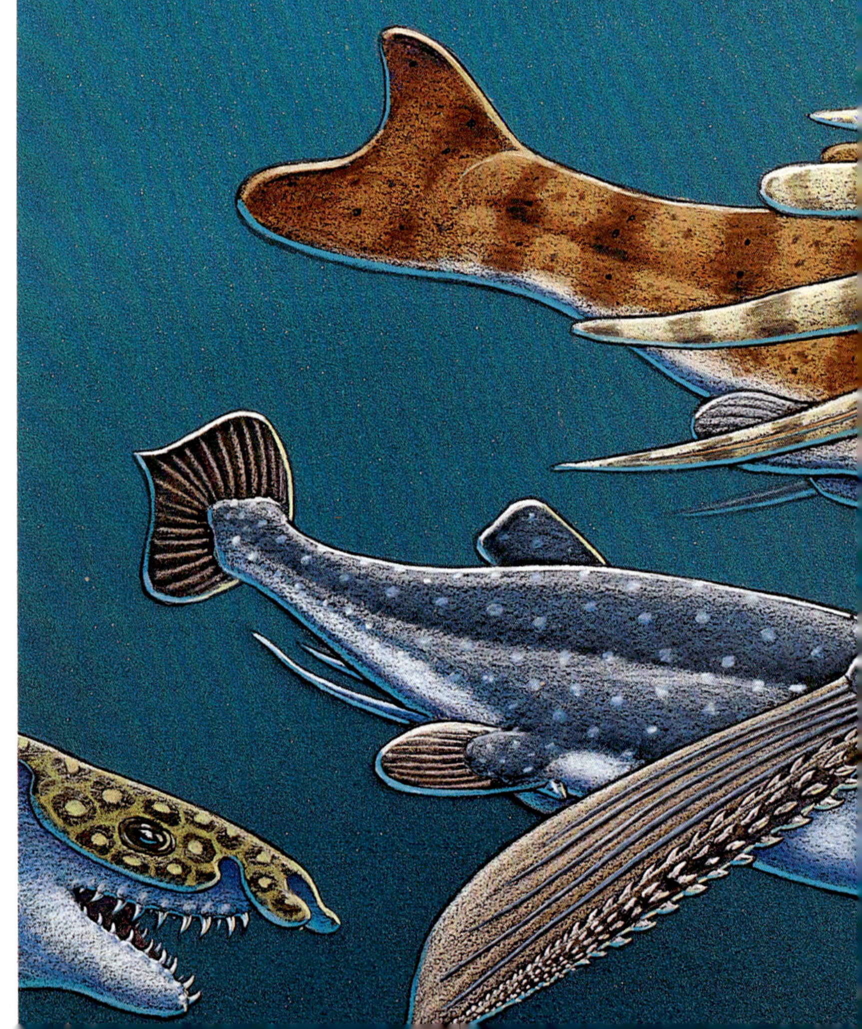

2001, colored pencil and acrylic paint on paper, 22 in. × 28 in.

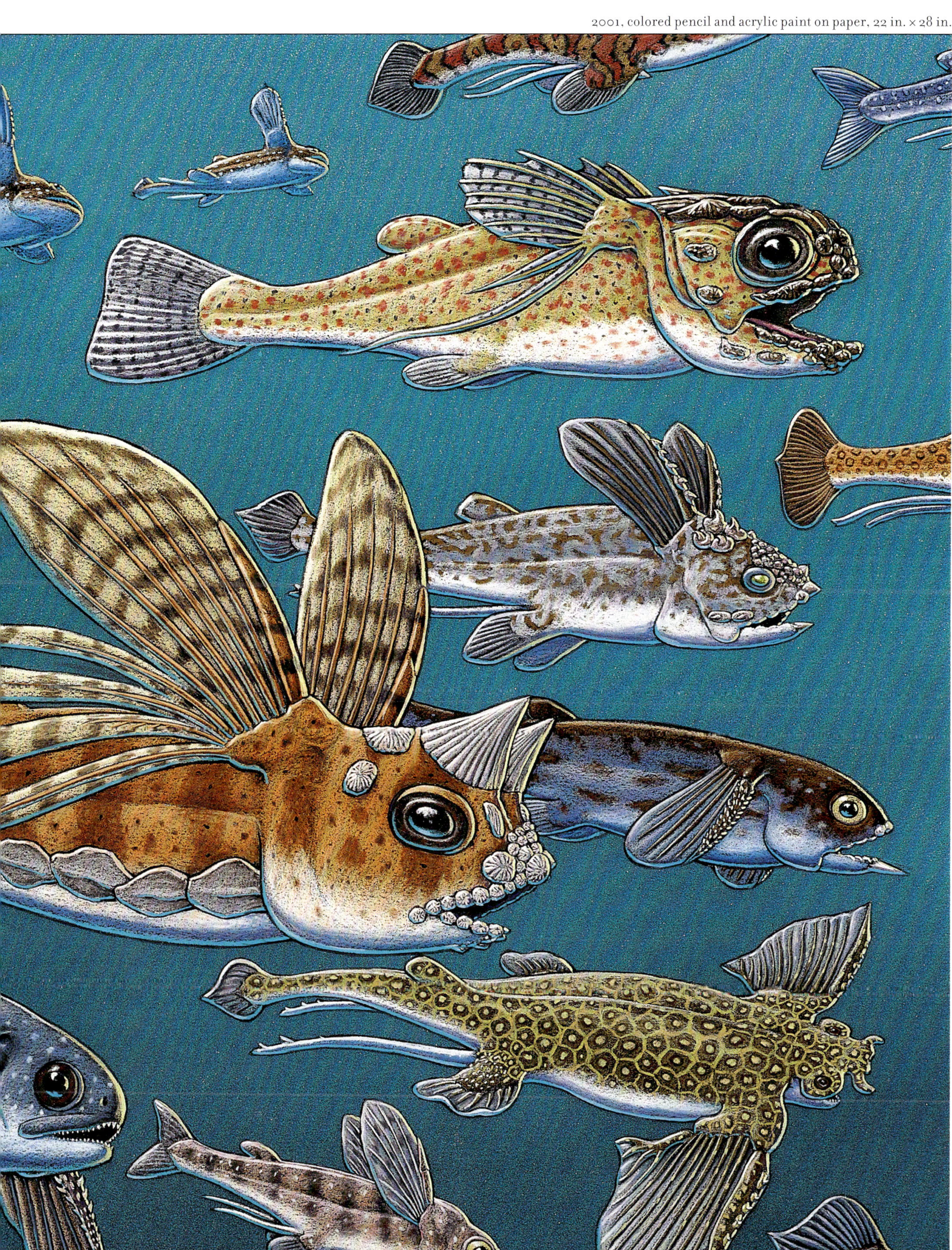

XENACANTH SHARK

2001, colored pencil on paper, 22 in. × 28 in.

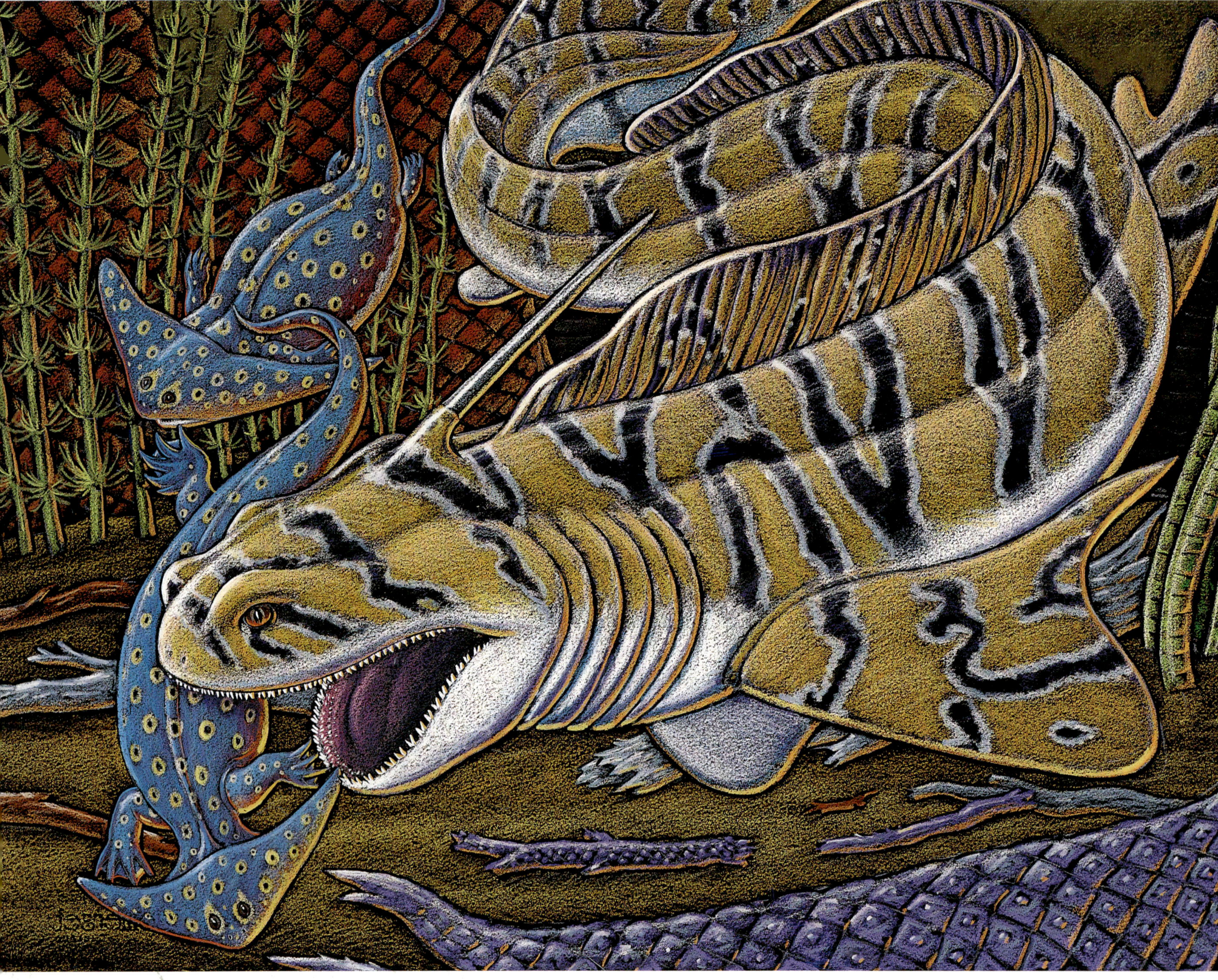

Xenacanths were large eel-shaped sharks that lived in swamps and rivers around the world. They were sinuous, snake-headed ambush predators, and nearly every species sported wicked, rearward-facing head spines. This is a Permian-period species from Texas called *Orthacanthus*, shown ambushing a wandering pair of amphibians called *Diplocaulus*.

SKATES ON ICE

1991, pastel on paper, 22 in. × 15 in.

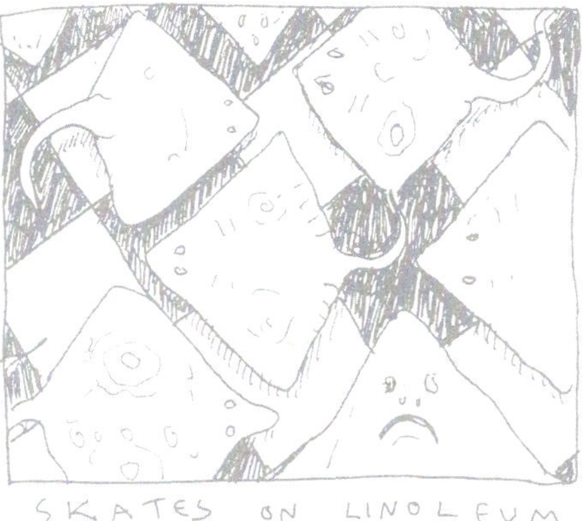

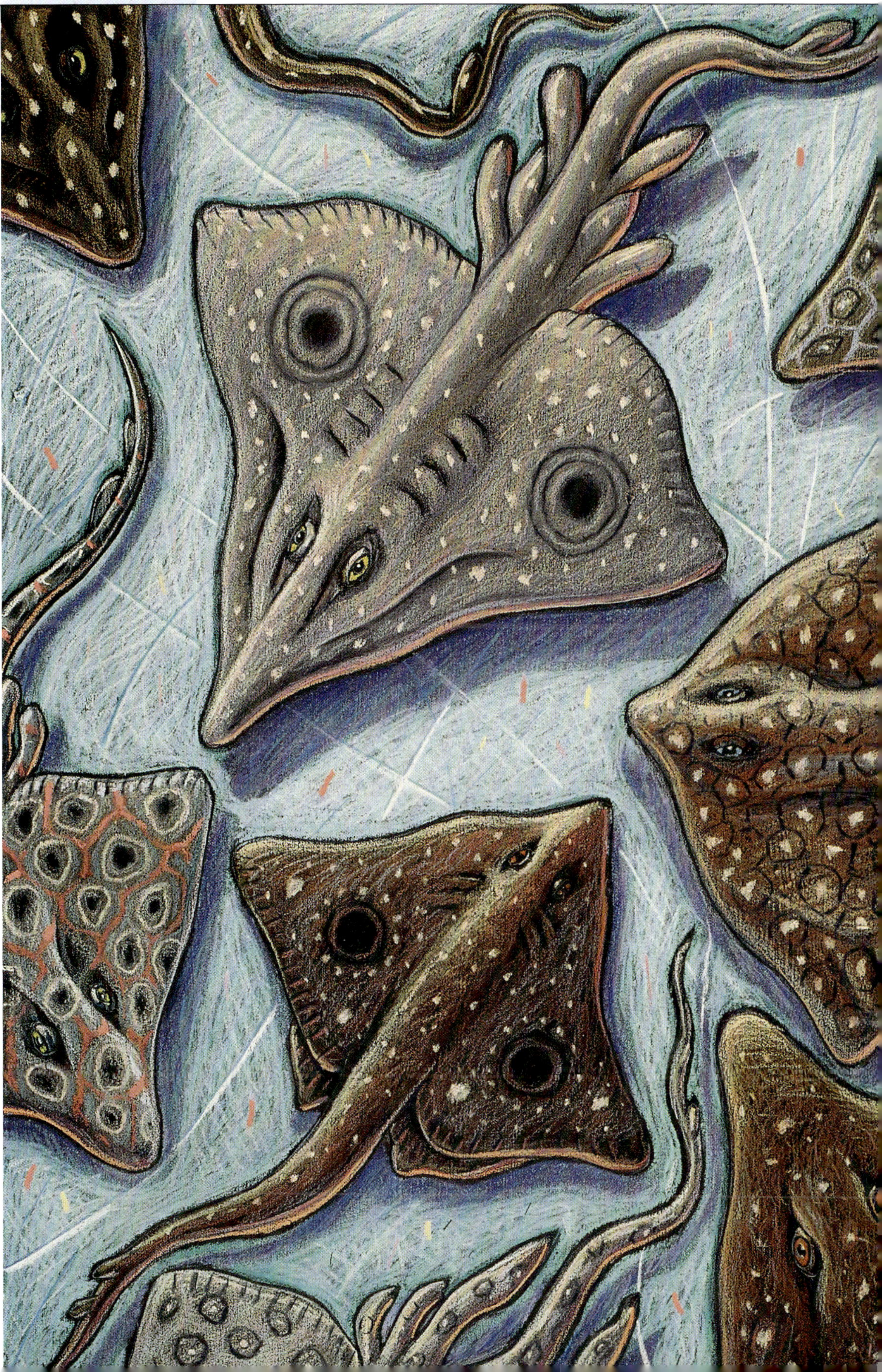

Skates and rays are members of the modern shark family. Like all sharks, they have a cartilaginous skeleton, gill slits, and electro-sensing capabilities, yet their bodies have evolved into an elegant flattened triangular shape. I like to call skates and rays "pancake sharks," a descriptive term coined by Dr. Sonny Gruber, a shark expert from Florida.

This early drawing of an obvious pun contains an error that bugs me now that I know better: I drew four gill slits on the top side of the animal at the top center. In fact, all skates have five pairs of gill slits, on their bottom sides. I live and learn. Many of my drawings show my scientific learning curve in action.

One afternoon in the spring of 1993 when we were working on *Planet Ocean*, Brad and I toured the extensive fossil collections of the Los Angeles County Museum of Natural History with our guide and mentor, Dr. J. D. Stewart. We spent quite a bit of time in a dank basement corner rummaging through rocks while J. D. explained some of the choice specimens. As we were leaving the room, I spotted a large boulder on the floor with a coiling spiral on its top surface. I thought at first it was an ammonite (an extinct relative of octopus and squid), which struck me as strange since we were in the vertebrate collection and ammonites don't have backbones. J. D. chuckled and said, "Look closer. It's a coil of shark teeth, and it's blown the minds of paleontologists for over a hundred years." Transfixed, I knew right then that figuring out this amazing paleo-puzzle would be one of my life's missions.

I dove into the literature and learned that a hapless Russian paleo-ichthyologist named Karpinsky spent decades in the late 1800s trying to figure out the position of the whorl of teeth on the living fish. His attempts were pretty hilarious. In several of his drawings the whorl emerges from the back of the beast, dangles off the upper snout, and even hangs from the tail. J. D. told me that one of the few people currently working on Paleozoic sharks was Dr. Rainer Zangerl, a retired shark expert from the Field Museum in Chicago. I called Dr. Z. the day I got back to my studio in Alaska and began a decade-long correspondence with this fascinating, wonderful scientist. The trail also led me to Dr. Richard Lund in New Jersey and Dr. Svend Bendix-Almgreen in Copenhagen (whom I surprised one night by calling him while he was eating dinner).

No scientific work had been done on this fascinating creature since 1964, when Bendix-Almgreen published a paper on fossilized whorls from Idaho, and no artist had ever come up with a believable representation of it. I plunged in headlong, running up huge long-distance phone bills in the process. Since that day in 1993 I've built models of whorl-tooth sharks and their relatives, scrambled across mine tailings looking for fossils in Idaho, studied all kinds of shark teeth, and even appeared on the Discovery Channel talking about my obsession. It's bordered on the absurd at times, but meanwhile I've become a bit of an authority on those weird ancient sharks, and the search has taught me a lot about my own perspective on the wonderful possibilities in nature.

This is the only animal in the history of life to cheat the tooth fairy. Its lower teeth were retained in its jaw in a wickedly bizarre spiral; as new, larger teeth erupted from the back of the mouth and moved forward, the older ones ratcheted down into the spiral. The upper jaw contained thousands of small crushing teeth. Basically, this was a swimming chopping block with a slicing tool. A slice 'n' dice shark.

One key to understanding the vexing whorl was the basic dental truth that teeth don't grow in size: they come in fully formed. When I first met Dr. Zangerl, he told me the oldest teeth, at the center of the whorl, were also the smallest teeth. At the time, I didn't quite get what that meant. Years later, though, I was having lunch with J. D. Stewart and Carl Ferraris in San Francisco, and they mentioned that baby sharks have tiny teeth. I almost wept as it all fell into place. Of course: the smallest teeth came into the mouth of the young shark first, with their successors growing increasingly larger, and they're all in the whorl if the shark doesn't shed them.

There is some disagreement about where this critter falls on the family tree of sharks. One camp puts it on the ratfish side of the cartilaginous fish tree, and the other puts it on the shark side. Since the only evidence is fossils of teeth and a few cranium/rostrum fragments, we are free to speculate in the scientific void and make educated guesses. I believe my *Helicoprion* is a fairly plausible reconstruction, but who knows. Perhaps a new fossil discovery will prove my attempt as laughable as Karpinsky's.

2000, colored pencil on paper, 22 in. × 28 in.

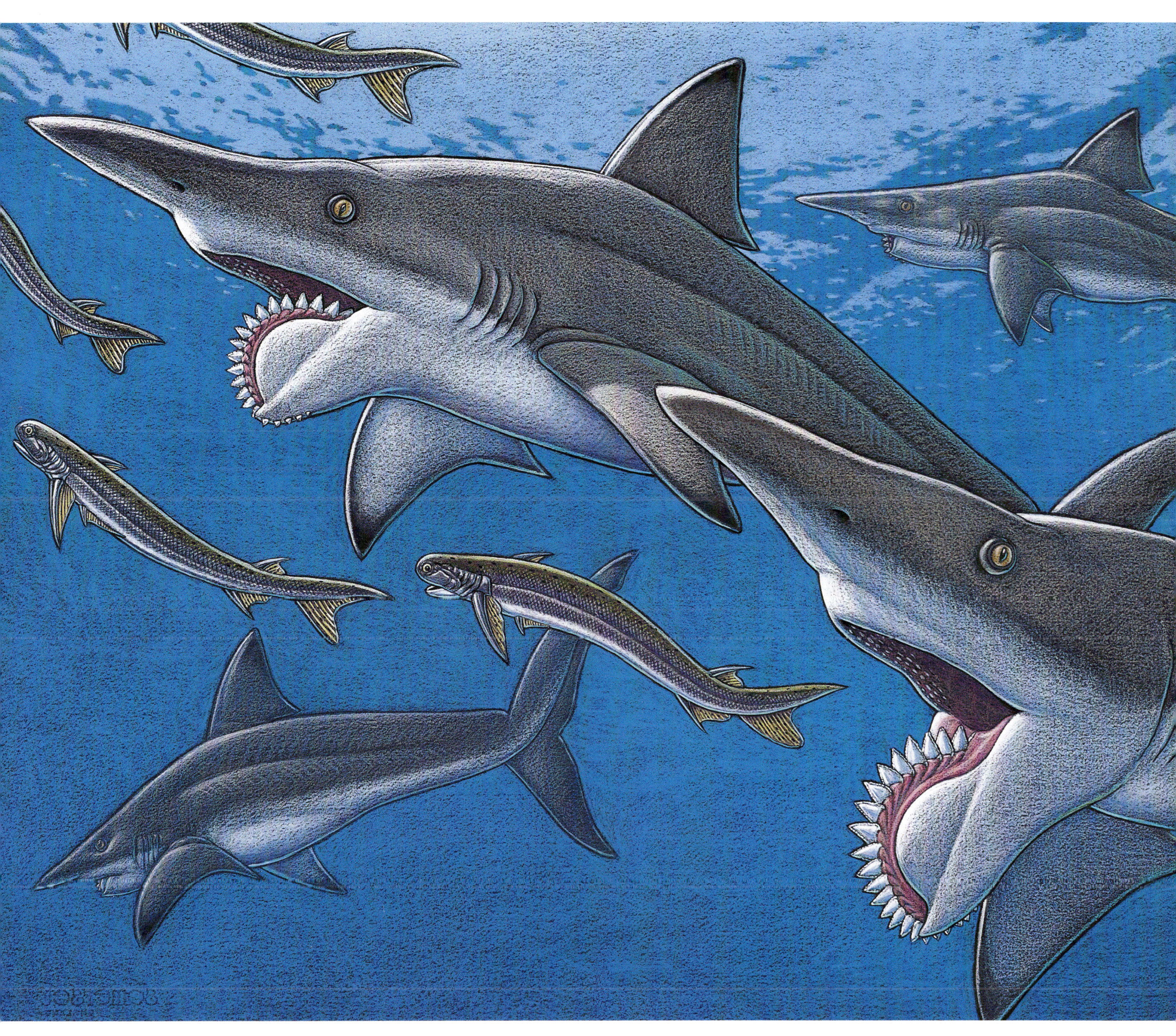

2001, colored pencil on paper, 7 in. × 9 in.

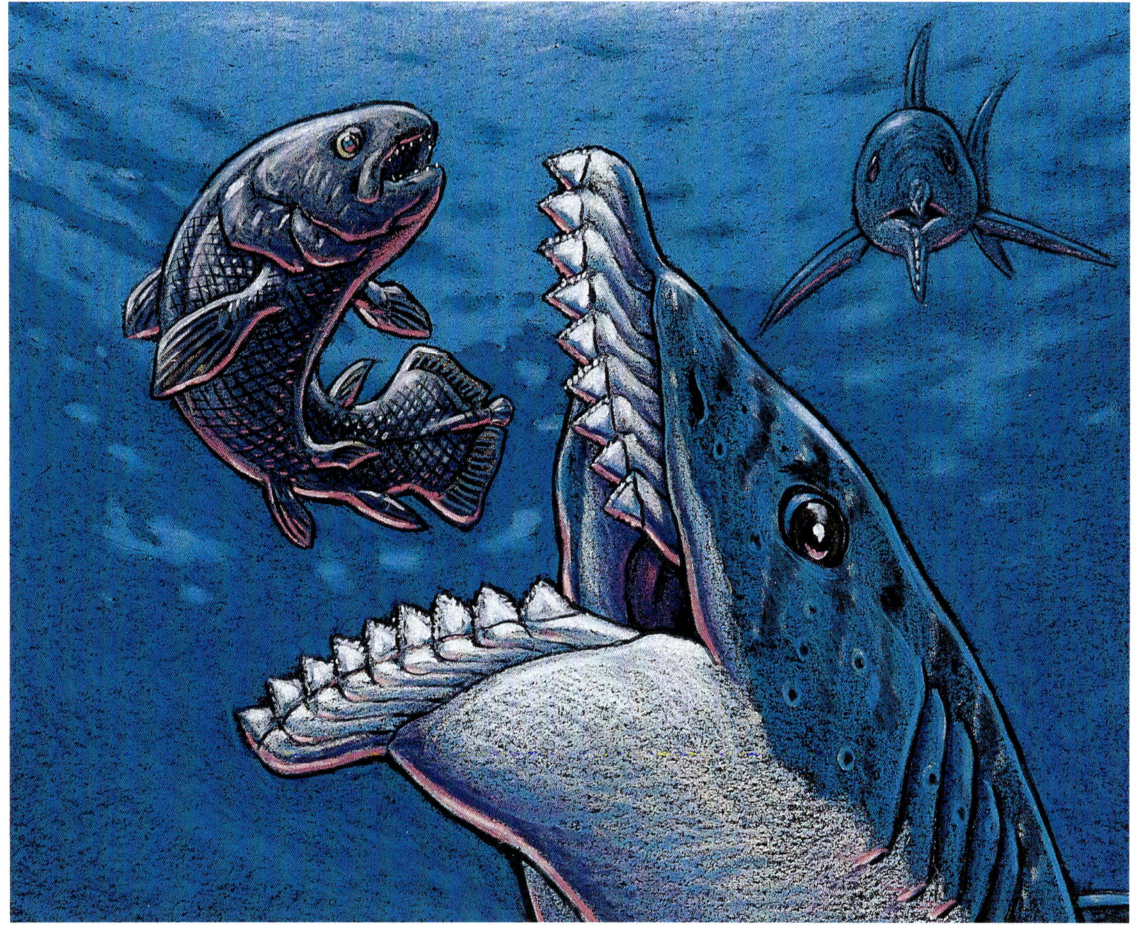

ONE OF my real joys over the last decade has been having the opportunity to make the first plausible reconstruction of a long-extinct animal. It's like being the first person in the world to see that creature. This drawing falls into that category. It's a Paleozoic shark named *Edestus giganteus* or, as I call it, the giant scissor-toothed shark. These malevolent-looking creatures must have been very adept at shearing their prey apart. From the size of the teeth of known fossils, I estimate that the shark was well over twenty feet long. The remarkable thing about this gargantuan species is that only one partial fossil of it has ever been found. It was unearthed in 1888, and no one had ever attempted to depict it until I did.

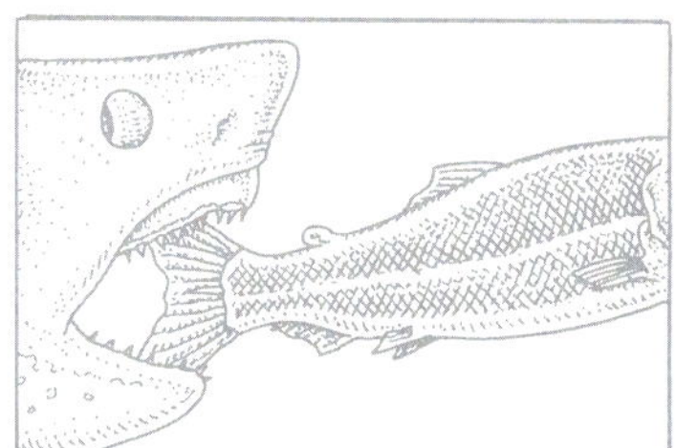

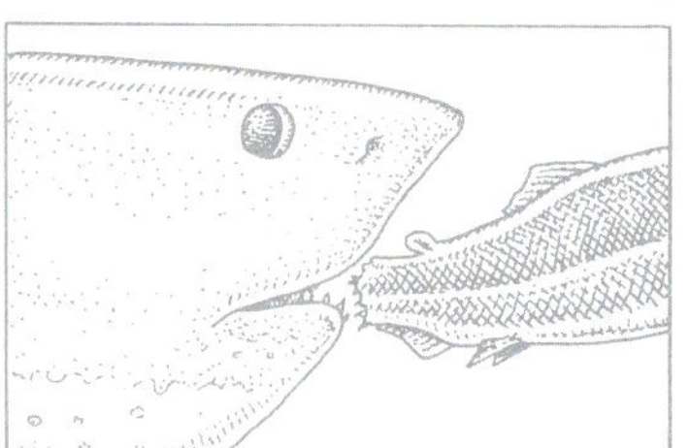

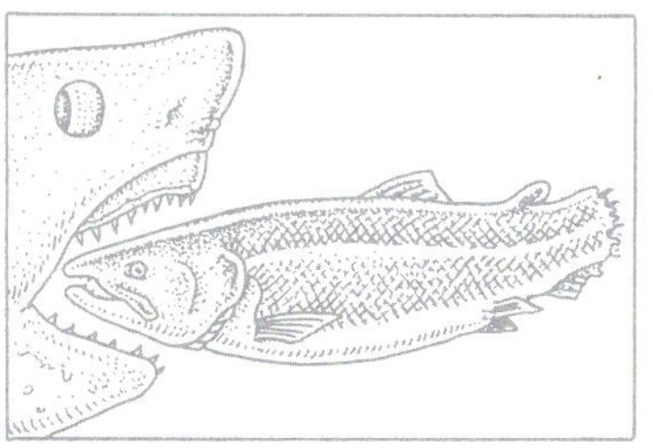

2001, colored pencil on paper, 7 in. × 9 in.

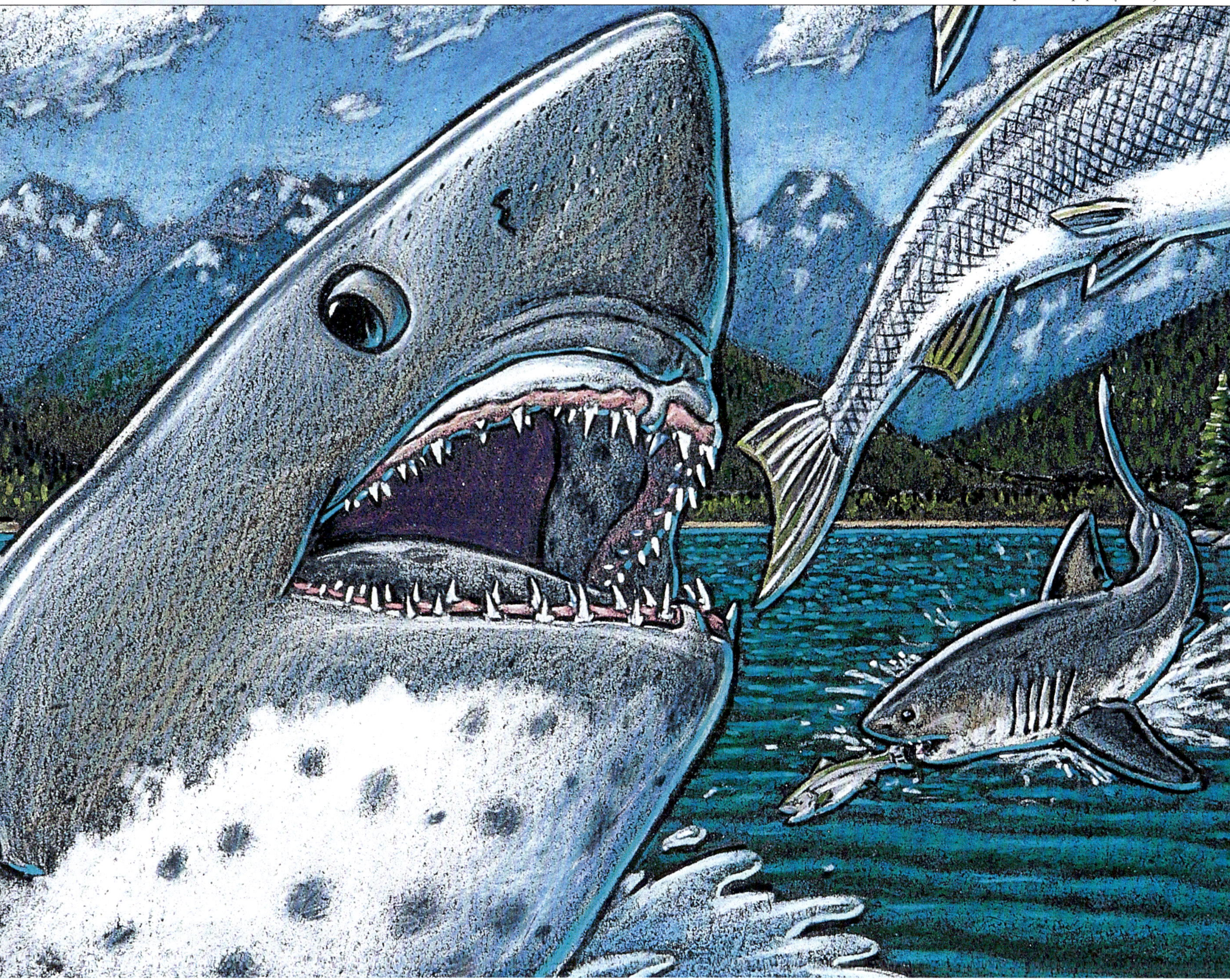

Salmon sharks converge on the coast of Alaska by the thousands in late summer to feed on returning schools of salmon. They're extremely agile and fast—they sometimes leap clear out of the water in hot pursuit of their prey—and are closely related to their bigger cousins the great whites. I was fortunate enough to go out on a National Marine Fisheries Service research trip in the summer of 2001 to help place satellite tags on salmon sharks near Cordova, Alaska. At times the water boiled and exploded around us as swarms of sharks attacked salmon near the boat.

2000, colored pencil on paper, 22 in. × 28 in.

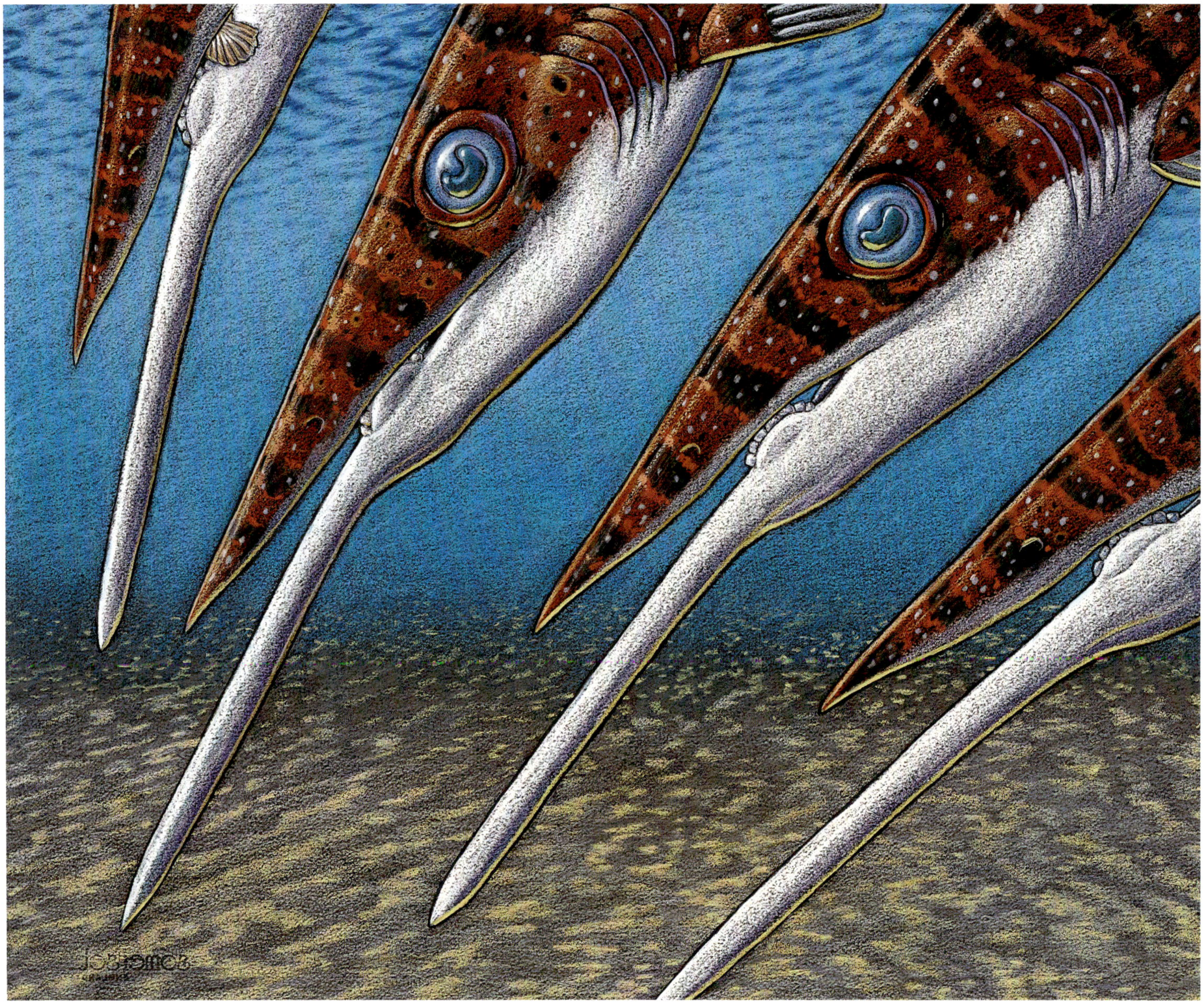

Ornithoprion was first described by Dr. Zangerl in 1966 from a fossil he found in the black shale of Indiana and Illinois. This bizarre little shark was related to *Helicoprion*, but instead of flesh-shearing teeth it evolved blunt, crushing teeth in a half whorl in its lower jaw. The odd-looking probe on its lower jaw was probably used to root around in sand and mud in pursuit of hard-shelled prey.

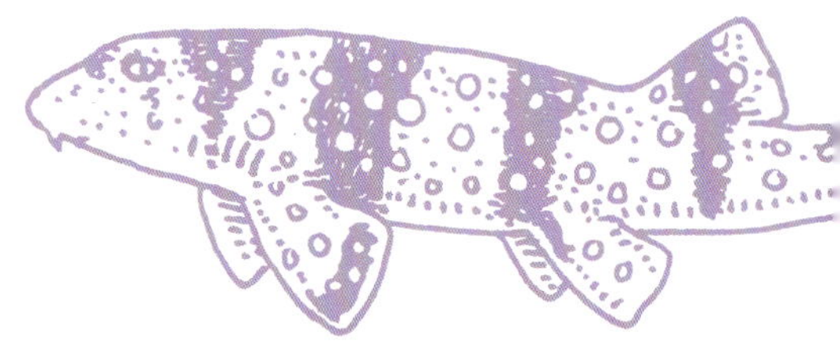

PETALODONT SHARK

2000, colored pencil on paper, 22 in. × 28 in.

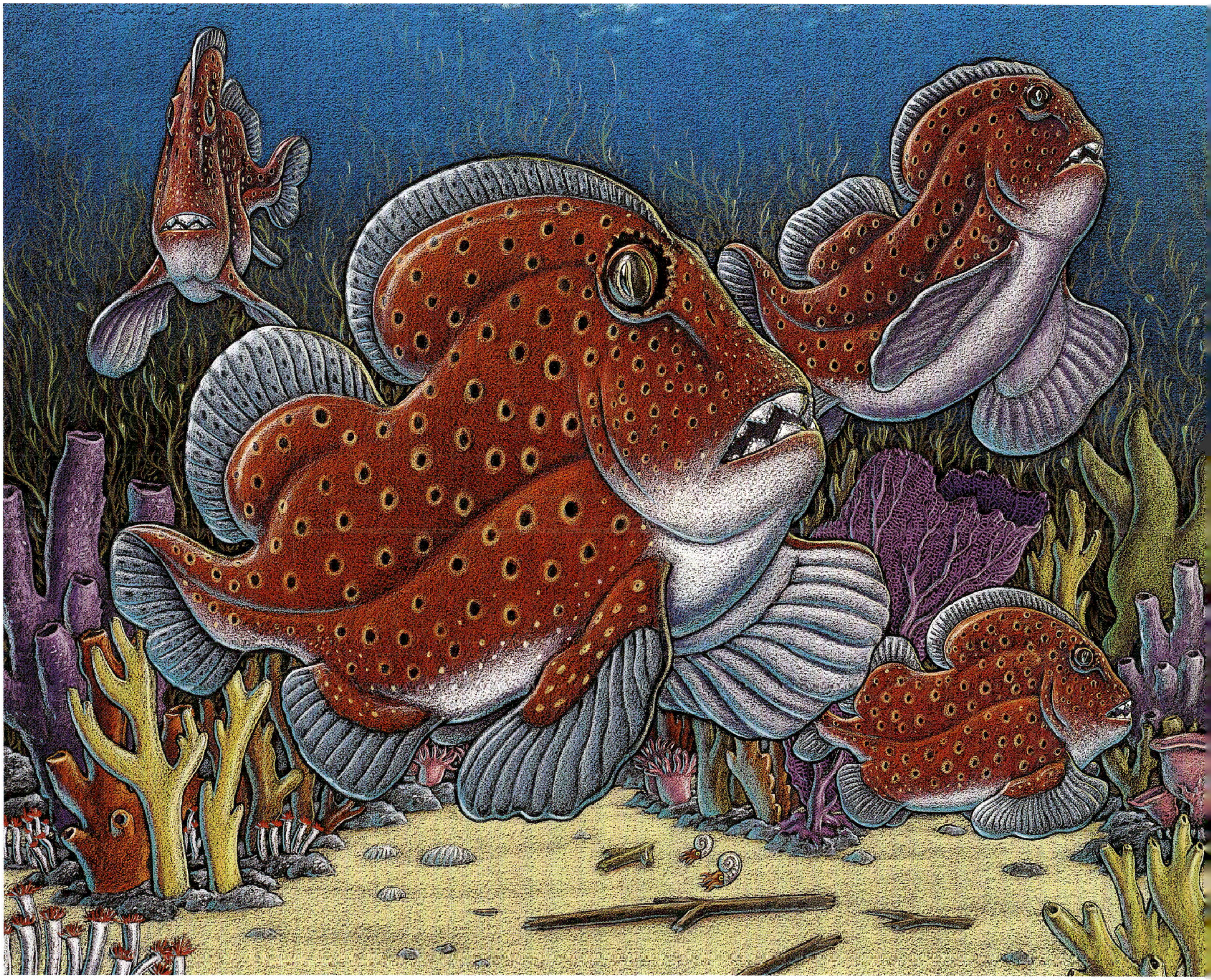

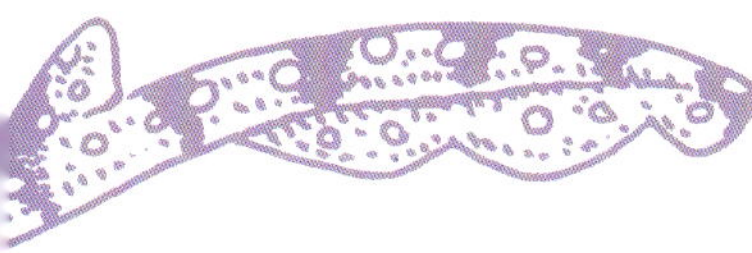

PETALODONTS ARE an entire order of extinct shark relatives, with hundreds of species known only by their petal-shaped teeth. Recently, more complete fossils have been found that divulge their funky body appearance. This buck-toothed beauty is named *Belantsea* and was discovered by Dr. Richard Lund in the famous paleo-shark fossil glory hole known as Bear Gulch, in Montana.

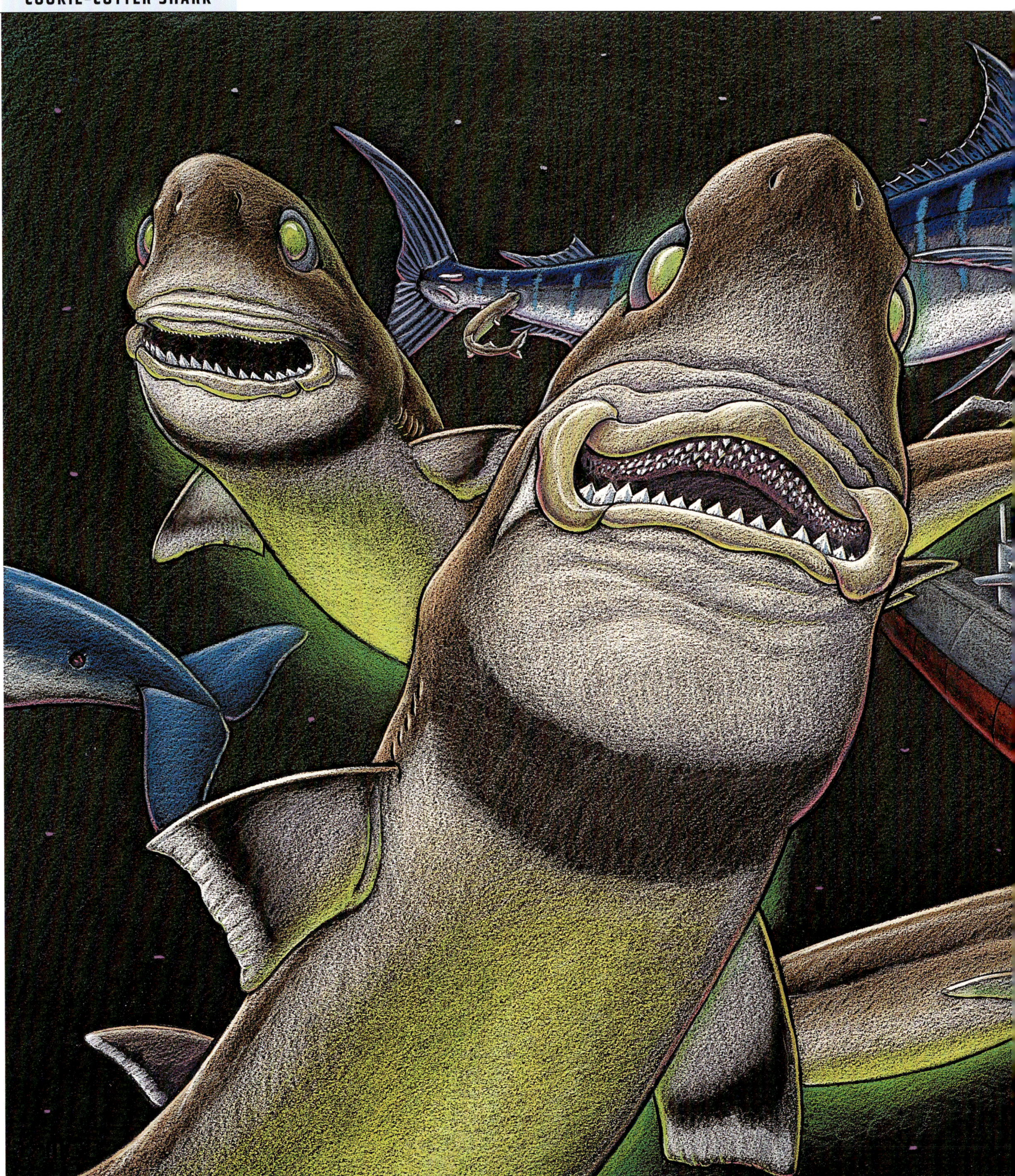

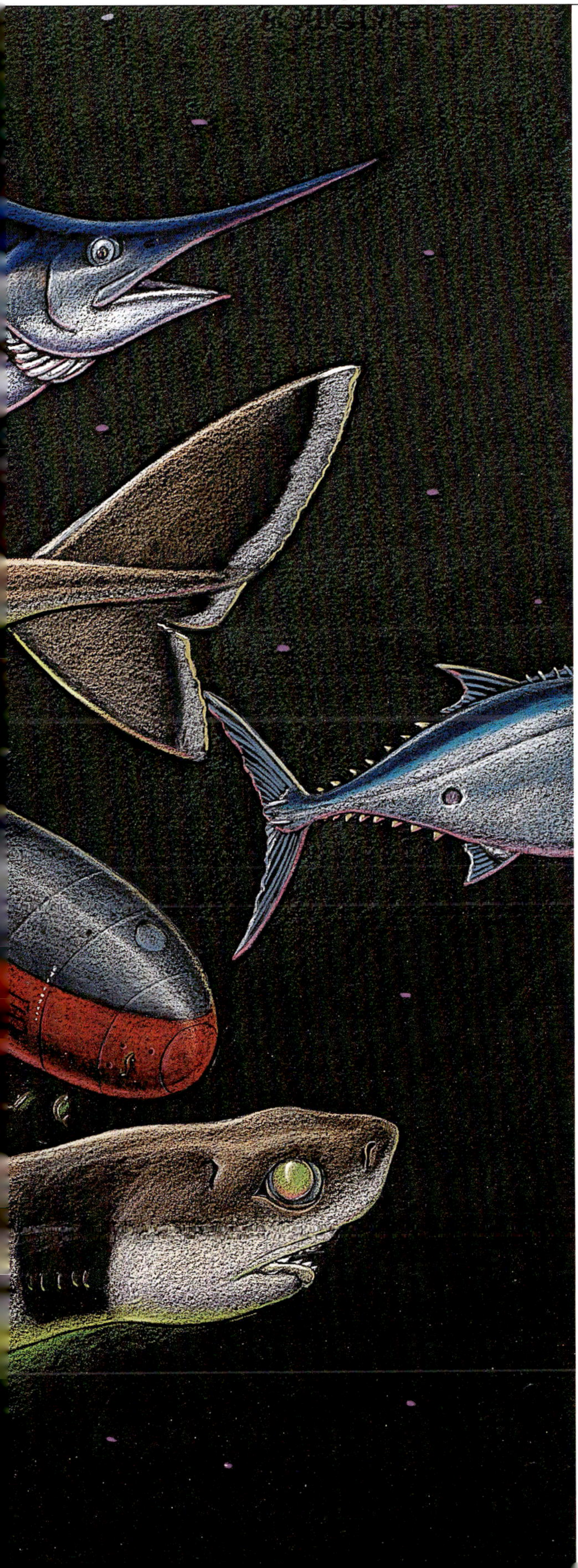

2001, colored pencil on paper, 22 in. × 28 in.

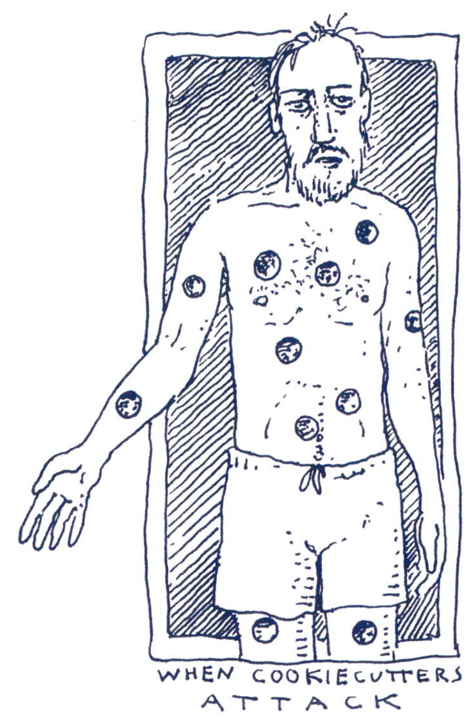

THIS LITTLE modern shark is a particular favorite of kids that have read *Sharkabet*. It grows to a mere eighteen inches but has eyes and a belly that glow in the dark, an enormous pair of lips, and a mouthful of razor-sharp teeth. Known formally as *Isistius brasiliensis*, it is notorious for taking cookie-sized bites out of such varied prey as dolphins, tuna, marlins, and even nuclear submarines. How these slow-moving sharks manage to do that is still somewhat of a mystery, but the prevailing theory is that they use their bioluminescence to lure their fast-swimming victims within biting range.

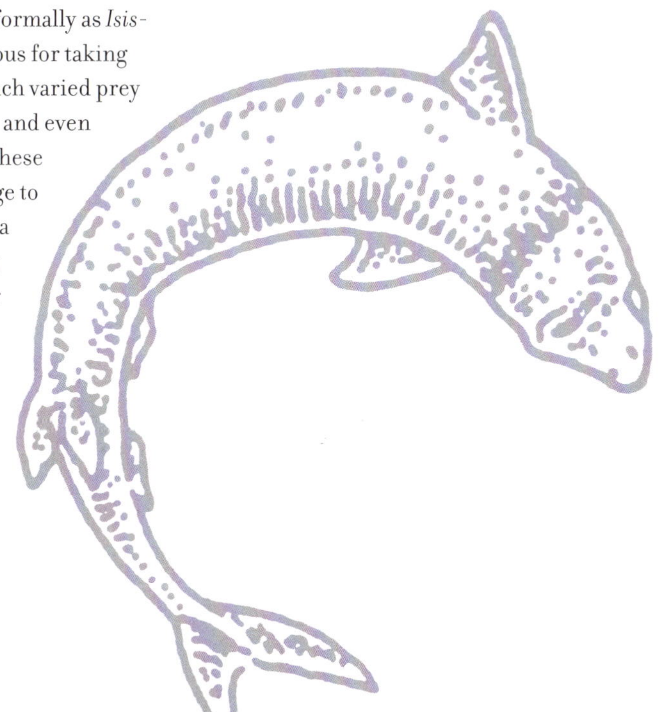

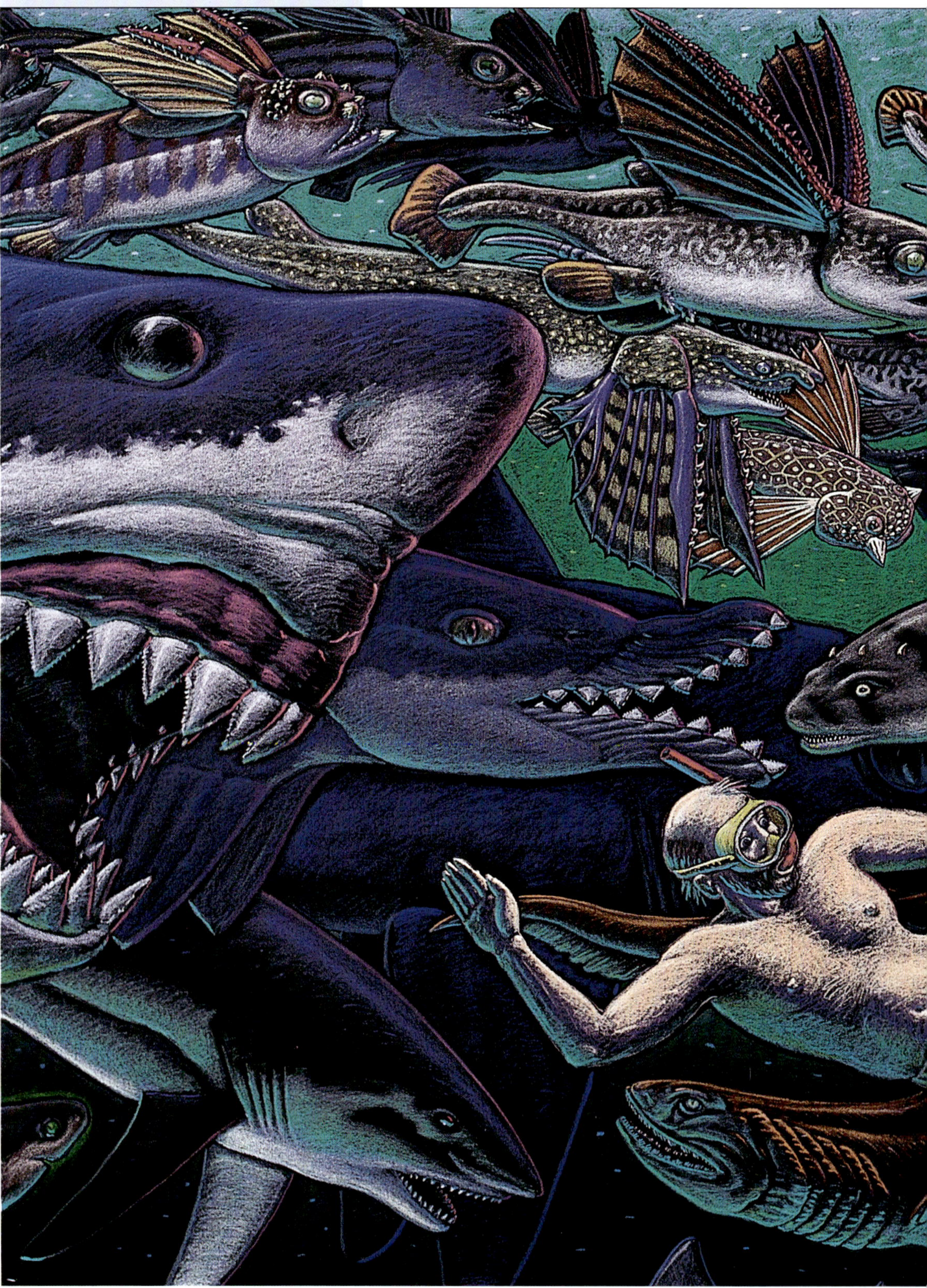

THIS IS a twelve-foot-long drawing of a nervous "deep time" snorkeler surrounded by living and extinct sharks and their relatives.

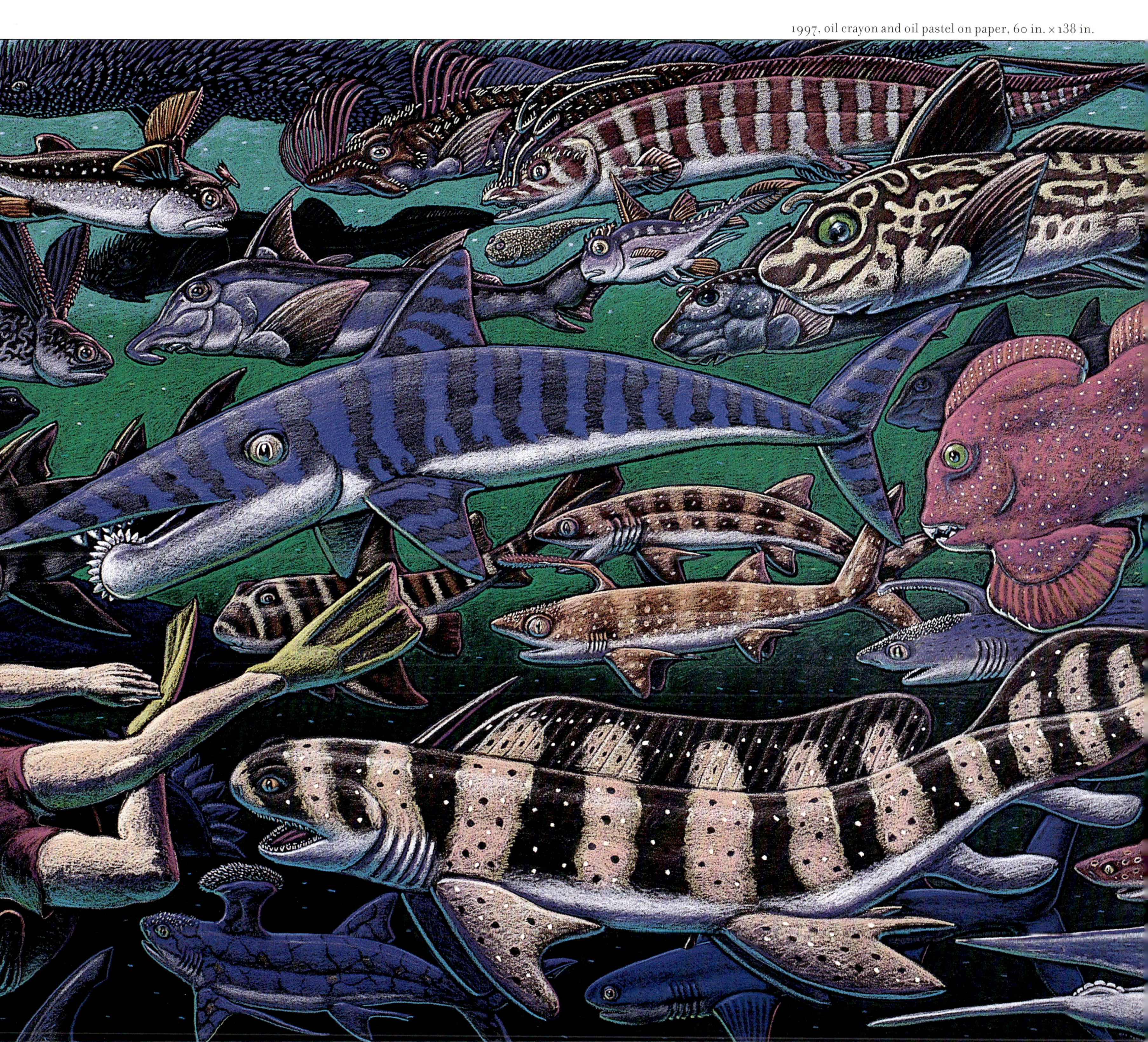

1997, oil crayon and oil pastel on paper, 60 in. × 138 in.

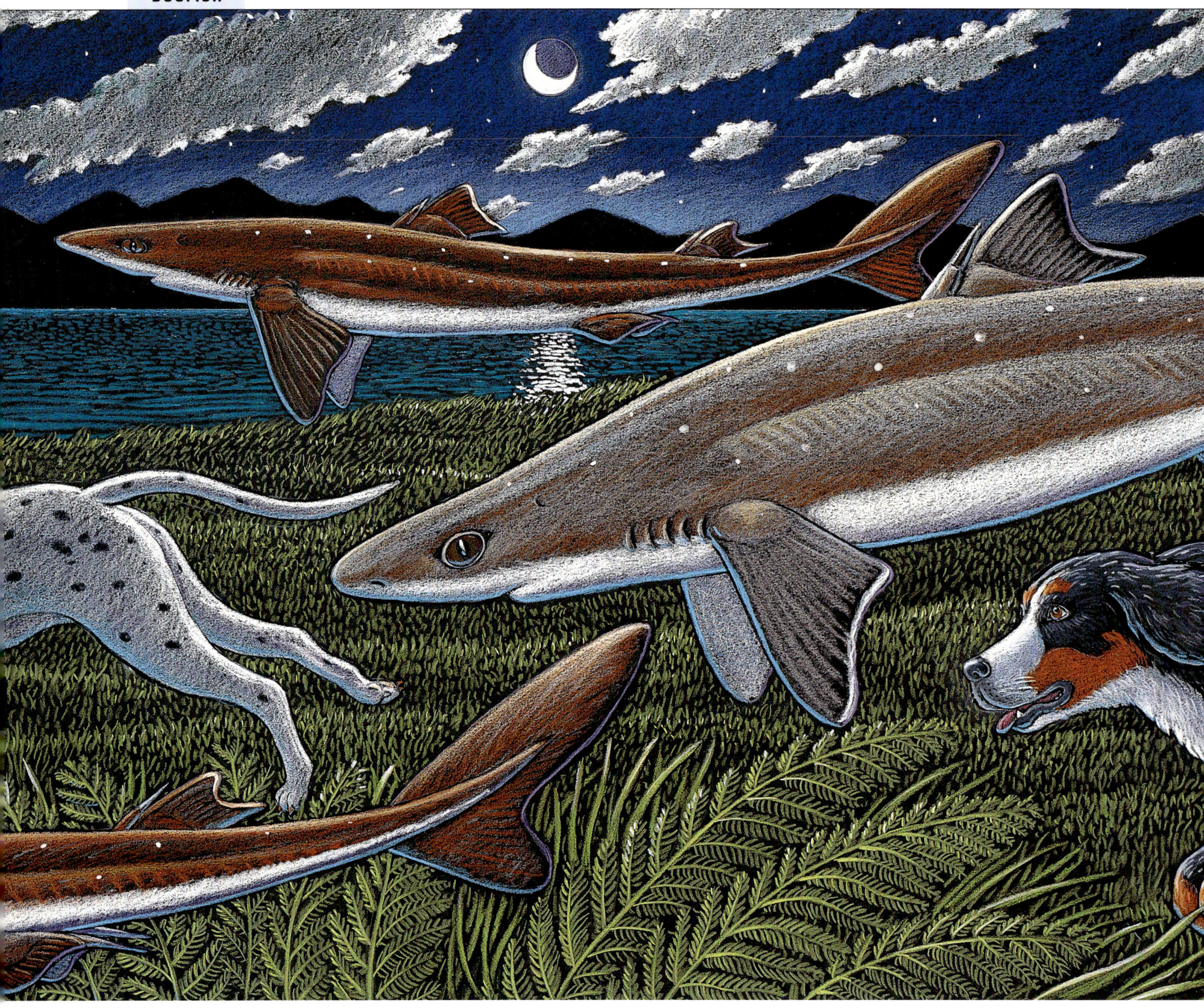

2001, mixed media on paper, 22 in. × 55 in.

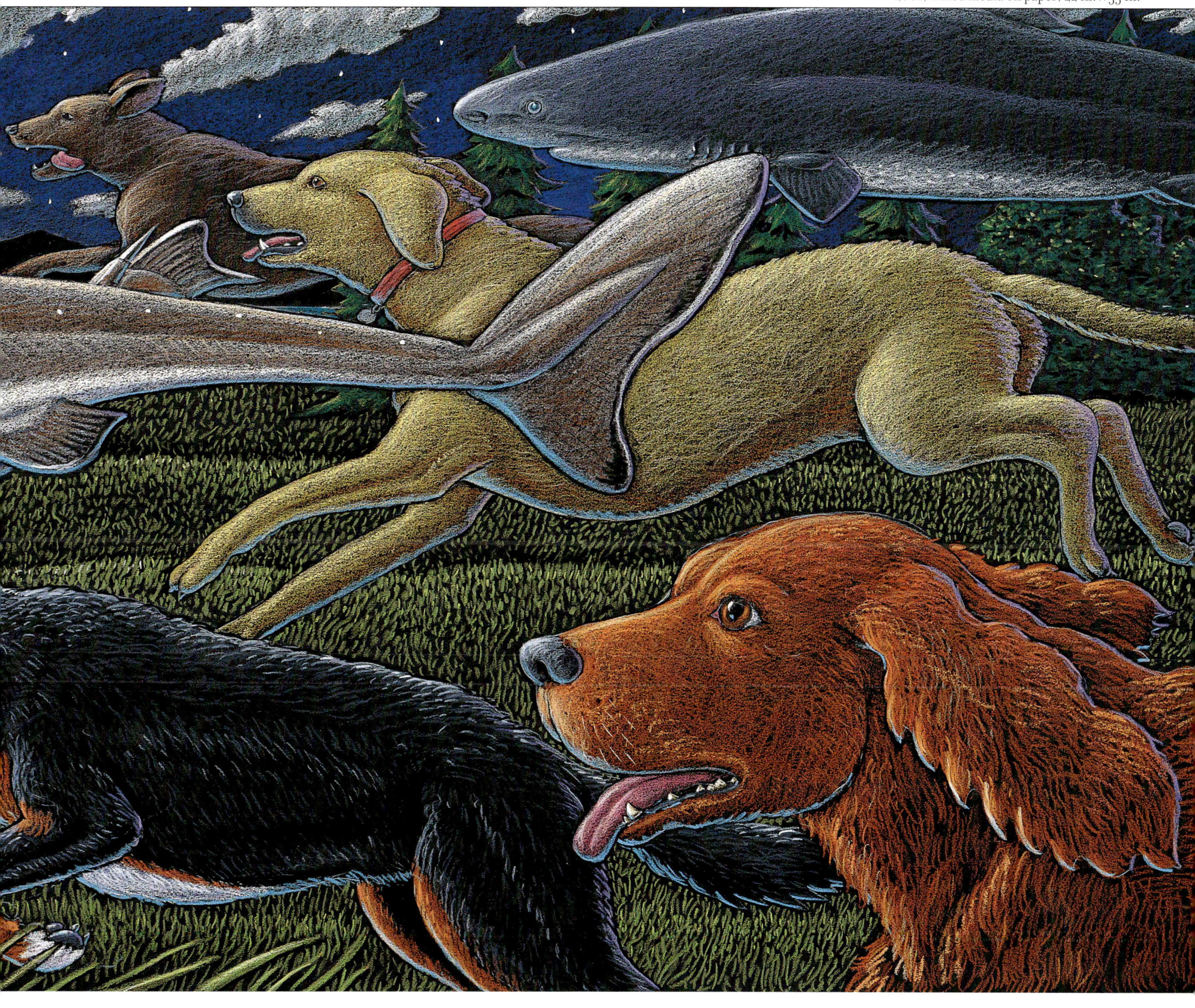

Dogfish is a name for an entire family of sharks over seventy species strong. I've drawn two types of dogfish in this image: the world's most common shark, the spiny dogfish; and, lurking in the background, the enormous Pacific sleeper shark. Dogfish earned their name because they travel in packs like dogs.

ILLUSTRATION CREDITS

The line drawings and marginalia scattered throughout this book are from the small mountain of sketchbooks that Ray Troll has been filling for over thirty years.

p. 4 Ketchikan, view of Water Street; photo by Ray Troll

p. 5 Docks of Ketchikan; photo by Ray Troll

p. 5 Fish processing plant; photo by Ray Troll

p. 6 Chief Johnson Tlingit totem pole, carved by Israel Shotridge; photo by Ray Troll

p. 7 Ray Troll and plastic red snapper; photo by Clark Mishler

p. 10 Ray Troll and mural *Kings*; photo by Normand Dupre, Thomas Street Imaging

p. 13 Ray Troll and saber-toothed salmon; photo by Richard Grost (www.Richardgrost.com)

p. 14 Ray Troll with pickled specimens; photo by Clark Mishler

pp. 20–21 *The Battle of Old Sitka, June 1802*, private collection; photo by John Wolon

p. 22 *Fog Woman*, private collection; photo by Normand Dupre, Thomas Street Imaging

p. 23 *Dance of the Fish Charmers*, private collection; photo by Normand Dupre, Thomas Street Imaging

p. 24 *Midnight Ritual*, collection of Wahluke School District, Mattawa, WA; photo by Normand Dupre, Thomas Street Imaging

p. 25 *Rain on the Parade*, collection of Alaska State Museum, Juneau, AK; photo by Joe Manfredini

pp. 26–27 *North Pacific Marine Life*, collection of Kodiak High School, Kodiak, AK; photo by Joe Manfredini

p. 28 *Creek Street*, private collection; photo by Normand Dupre, Thomas Street Imaging

p. 29 *The Ballad of Old One Eye*, private collection; photo by Normand Dupre, Thomas Street Imaging

p. 30 *Ketchikan Still Life*, private collection; photo by Normand Dupre, Thomas Street Imaging

p. 31 *Sockeye Warriors*, collection of Lake Stevens School District, Lake Stevens, WA; photo by Normand Dupre, Thomas Street Imaging

pp. 32–33 *Midnight Run*, collection of Point Higgins Elementary School, Ketchikan, AK; photo by Joe Manfredini

p. 34 *Alaska, the Last Great Adventure*, private collection; photo by Normand Dupre, Thomas Street Imaging

p. 35 *The Kingfisher, the Raven, and the Half Moon Run*, private collection; photo by Joe Manfredini

p. 36 *Eagle Bait*, private collection; photo by Normand Dupre, Thomas Street Imaging

p. 37 *The Avenging Halibut*, private collection; photo by Normand Dupre, Thomas Street Imaging

p. 40 *Know Your Salmon, Bub*, private collection; photo by Normand Dupre, Thomas Street Imaging

p. 41 *Fishscape*, private collection; photo by Normand Dupre, Thomas Street Imaging

pp. 42–43 *Gamefish*, private collection; photo by Normand Dupre, Thomas Street Imaging

p. 44 *Bottom Fish of the North Pacific*, private collection; photo by Normand Dupre, Thomas Street Imaging

p. 45 *Kings and Queens*, private collection; photo by Normand Dupre, Thomas Street Imaging

pp. 46–47 *Fish Wars*, collection of Alaska State Council on the Arts Art Bank, Anchorage, AK; photo by Normand Dupre, Thomas Street Imaging

p. 48 *Rockfish*, private collection; photo by Normand Dupre, Thomas Street Imaging

p. 49 *Twist and Trout*, private collection; photo by Normand Dupre, Thomas Street Imaging

p. 50 *Weapons of Bass Destruction*, private collection; photo by Normand Dupre, Thomas Street Imaging

p. 51 *Dream of the Fisherman*, private collection; photo by Normand Dupre, Thomas Street Imaging

p. 52 *Metalhead*, private collection; photo by Normand Dupre, Thomas Street Imaging

p. 53 *Rebel without a Cod*, private collection; photo by Normand Dupre, Thomas Street Imaging

p. 54 *Cutthroat Businessmen*, private collection; photo by Normand Dupre, Thomas Street Imaging

p. 55 *Bassackwards*, private collection; photo by Normand Dupre, Thomas Street Imaging

p. 56 *Life's a Fish and Then You Fry*, private collection; photo by Normand Dupre, Thomas Street Imaging

p. 57 *Ain't No Nookie Like Chinookie*, collection of the Anchorage Museum of History and Art, Anchorage, AK; photo by Normand Dupre, Thomas Street Imaging

p. 58 *Hell's Anglers*, private collection; photo by Normand Dupre, Thomas Street Imaging

p. 59 *Spawn Till You Die*, private collection; photo by Normand Dupre, Thomas Street Imaging

pp. 60–61 *Humpies from Hell*, private collection; photo by Normand Dupre, Thomas Street Imaging

p. 62 *Fish Head*, private collection; photo by Normand Dupre, Thomas Street Imaging

p. 63 *Sometimes Late at Night . . .*, private collection; photo by Normand Dupre, Thomas Street Imaging

p. 64 *There Is No Free Lunch*, private collection; photo by Normand Dupre, Thomas Street Imaging

p. 65 *Fish Worship*, private collection; photo by Normand Dupre, Thomas Street Imaging

pp. 66–67 *Stream of Consciousness*, collection of the Anchorage Museum of History and Art, Anchorage, AK; photo by Normand Dupre, Thomas Street Imaging

pp. 68–69 *Rapture of the Deep*, private collection; photo by Normand Dupre, Thomas Street Imaging

p. 72 *The Last View for Many a Cretaceous Fish*, private collection; photo by Normand Dupre, Thomas Street Imaging

p. 73 *Professor Williston's Last Supper*, private collection; photo by Normand Dupre, Thomas Street Imaging

p. 74 *Trilos by Fire*, collection of the California Academy of Sciences, San Francisco, CA; photo by Normand Dupre, Thomas Street Imaging

p. 75 *Hallucigenia*, private collection; photo by Normand Dupre, Thomas Street Imaging

p. 76 *The Lucky Fish Gets the Cheeseburger*, private collection; photo by Normand Dupre, Thomas Street Imaging

p. 77 *In the Midday Sun . . .*, private collection; photo by Normand Dupre, Thomas Street Imaging

pp. 78–79 *Night of the Giant Ammonites*, private collection; photo by Normand Dupre, Thomas Street Imaging

p. 80 *Leedsichthys, Bigger Than My House*, private collection; photo by Normand Dupre, Thomas Street Imaging

p. 81 *Trilobite Safari*, private collection; photo by Normand Dupre, Thomas Street Imaging

p. 82 *Xiphactinus, the Meanest Pescado*, private collection; photo by Normand Dupre, Thomas Street Imaging

p. 83 *Cretaceous Road Dream*, private collection; photo by Normand Dupre, Thomas Street Imaging

pp. 84–85 *Not in Kansas Anymore*, private collection; photo by Normand Dupre, Thomas Street Imaging; key drawn by Bill Nelson

pp. 86–87 *Planet Ocean*, private collection; photo by Normand Dupre, Thomas Street Imaging

p. 88 *The Way We Were*, private collection; photo by Normand Dupre, Thomas Street Imaging

p. 89 *From the Slime to the Ridiculous*, private collection; photo by Normand Dupre, Thomas Street Imaging

pp. 90–91 *Dancing to the Fossil Record*, private collection; photo by Normand Dupre, Thomas Street Imaging

p. 92 *Dancing to the Fossil Record* exhibit, Academy of Natural Sciences, Philadelphia; photo by Sean Duran

p. 93 "Night of the Giant Ammonites" installation. All rights reserved, Image Archives, Denver Museum of Nature & Science; photos by Rick Wicker

pp. 94–95 *Drivin' with Darwin*, private collection; photo by Normand Dupre, Thomas Street Imaging

p. 95 "Evolvo" installation. All rights reserved, Image Archives, Denver Museum of Nature & Science; photo by Rick Wicker

p. 98 Piranha; photo by Ray Troll

p. 99 *Vicious Fishes of the Amazon*, collection of the artist; photo by John Wolon

p. 100 Moacir Fortés with arowana; photo by Ray Troll

p. 100 Piraíba; photo by Ray Troll

pp. 101–103 *Fishes of Amazonia*, collcction of the artist; photo by John Wolon; key drawn by Bill Nelson

p. 104 *River Rays*, collection of the Anchorage Museum of History and Art, Anchorage, AK; photo by John Wolon

p. 105 *Remembering Gondwana*, collection of Dr. Kirk Johnson, Denver, CO; photo by John Wolon

p. 108 *Catfish, Ratfish*, collection of Burk's Café, Seattle, WA; photo by Joe Manfredini

p. 109 *Cardsharks*, private collection; photo by John Wolon

pp. 110–111 *A Ratfish Called Troll*, collection of the artist; photo by John Wolon

pp. 112–113 *Iniopterygians*, collection of the artist; photo by John Wolon

p. 114 *Xenacanth Shark*, collection of the artist; photo by John Wolon

p. 115 *Skates on Ice*, collection of Captain Strong Elementary School, Battleground, WA; photo by Normand Dupre, Thomas Street Imaging

p. 117 *Helicoprion*, collection of the artist; photo by John Wolon

p. 118 *Giant Scissor-Toothed Shark*, collection of the artist; photo by John Wolon

p. 119 *Salmon Sharks*, collection of the artist; photo by John Wolon

p. 120 *Ornithoprion*, collection of the artist; photo by John Wolon

p. 121 *Petalodont Shark*, collection of the artist; photo by John Wolon

pp. 122–123 *Cookie-Cutter Shark*, collection of the artist; photo by John Wolon

pp. 124–125 *Swimming with the Sharks*, collection of the Moss Landing Marine Laboratories, Moss Landing, CA; photo by Normand Dupre, Thomas Street Imaging

pp. 126–127 *Dogfish*, private collection; photo by John Wolon

ACKNOWLEDGMENTS

Many people contributed to bringing *Rapture of the Deep* into the world. We'd like to thank Milton Love for dreaming up the idea in the first place; Doris Kretschmer for believing in it, being a great ringleader, and making it happen; Nola Burger for her artful design skills; Rose Vekony and Lynn Meinhardt for cleaning up our language; Anne Canright (love that name!) for polishing up Ray's captions; and John Cronin for bringing the book into living color. Thanks also to Kate Thompson for sorting through the art; David James Duncan for his superb metaphysaesthetical foreword; John Lundberg for his support; Kes Woodward for going to bat for this book when we needed it; the photographers who graciously loaned their work; Greg Cailliet, Kerry Tremain, and Barbara Ramsey for housing us on our California road trip; the cast and crew at the Elkhorn Yacht Club and the Queen City Grill for their moral support; and the Frontier Room for the barbecue that launched us on our way.

Ray is especially indebted to all the collectors over the last thirty years who have bought his works, from T-shirts to over-sized murals; to the many scientists with whom he has collaborated in portraying weird and wonderful critters; and, of course, to the fish.

Our families—Michelle, Patrick, Corinna, Laara, Jonas, and Milo—as well as friends too numerous to name, kept us sane and happy as usual, and for them our gratitude knows no bounds.

Ray Troll and Brad Matsen

From his studio high above the Tongass Narrows in rainswept Ketchikan, Alaska, **RAY TROLL** draws and paints fishy images that migrate into museums, books, and magazines and onto T-shirts sold round the globe. His aquatic images are informed by the latest scientific discoveries, but Ray also brings a street-smart sensibility to the worlds of ichthyology and paleontology.

Troll was born in Corning, New York, in 1954, and raised as an Air Force brat in a family of six kids and in eleven different locales around the States and overseas. After earning a Bachelor of Arts degree from Bethany College in Lindsborg, Kansas, in 1977 and an M.F.A. in studio arts from Washington State University in 1981, he moved to Ketchikan in 1983 to spend a summer helping his sister start a seafood retail store. The fish store is long gone, but Ray is still there.

Troll's unique blend of art and science blossomed into his traveling exhibit *Dancing to the Fossil Record*, which opened at the California Academy of Sciences in San Francisco in 1995. The show included Ray's original drawings, gigantic fossils, tanks of live fish, murals, an original soundtrack, a dance floor, and an interactive computer installation. In 1997 the exhibit traveled to the Oregon Coast Aquarium in Newport, in 1998 to the Philadelphia Academy of Natural Sciences, and the tour ended in 1999 at the Denver Museum of Nature and Science. By that time it had grown to fourteen thousand square feet and had had several names.

Ray currently has a touring museum show based on his new children's book, *Sharkabet: A Sea of Sharks from A to Z*. It has been booked into the Science Museum of Minnesota in Minneapolis, the Anchorage

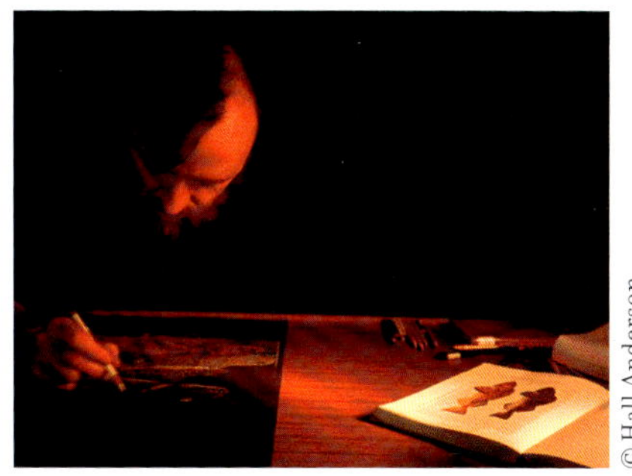
© Hall Anderson

Museum of History and Art, the Mesa (Arizona) Southwest Museum, the Tongass Historical Museum in Ketchikan, and the Museum of the Rockies in Bozeman, Montana. Ray is also working on another traveling exhibit, called *Amazon Voyage: Vicious Fishes and Other Riches*, with the Miami Museum of Science. In addition to *Sharkabet*, Ray has illustrated three books with author Brad Matsen—*Ray Troll's Shocking Fish Tales*; *Planet Ocean: A Story of Life, the Sea, and Dancing to the Fossil Record*; and *Raptors, Fossils, Fins, and Fangs*—and did the illustrations for *Life's a Fish and Then You Fry* by chef Randy Bayliss.

Ray is proud to be an honorary member of the Gilbert Ichthyological Society, the Guild of Natural Science Illustrators, and a lifetime member of S.P.O.O.F. (Society for the Protection of Old Fish).

Ray and his wife, Michelle, own and manage the Soho Coho Contemporary Art and Craft Gallery, located on a spawning stream in the former red-light district of Ketchikan. The gallery features Ray's own artwork, T-shirts, and fish juju, as well as original artwork by other local artists.

When not at the drawing board, Ray is hanging out with his two teenage kids, Corinna and Patrick. He is also an avid fossil-collecting rockhound and, of course, an ardent angler (with questionable fishing skills).

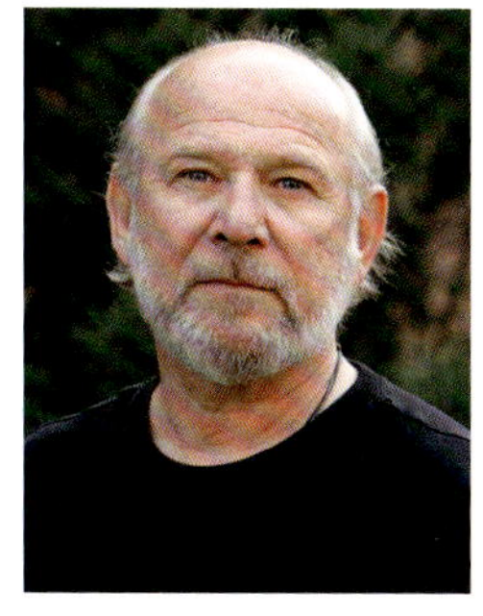
© Jonas Bendiksen

BRAD MATSEN has been writing about the sea and its creatures for twenty-five years in books, film scripts, and magazine articles. His books include *Ray Troll's Shocking Fish Tales*, *Planet Ocean*, *Fishing Up North*, and *Faces of Fishing*. His newest book, *Descent: The Heroic Discovery of the Abyss*, will be published in the fall of 2005. He was a creative producer on the *Shape of Life*, an eight-hour Sea Studios/National Geographic television series that aired on PBS in April 2002, and he wrote the accompanying book of the same name. He has written for *Mother Jones*, *Audubon*, *Nature*, and many other magazines; his 1999 coverage of depleted ocean resources in *Mother Jones*, "Blues in the Key of Sea," won the Project Censored Prize as one of that year's ten best stories. From 1985 to 1995 he was the Pacific editor of *National Fisherman* and a senior editor of *Seafood Business*. He lives in Seattle and New York.

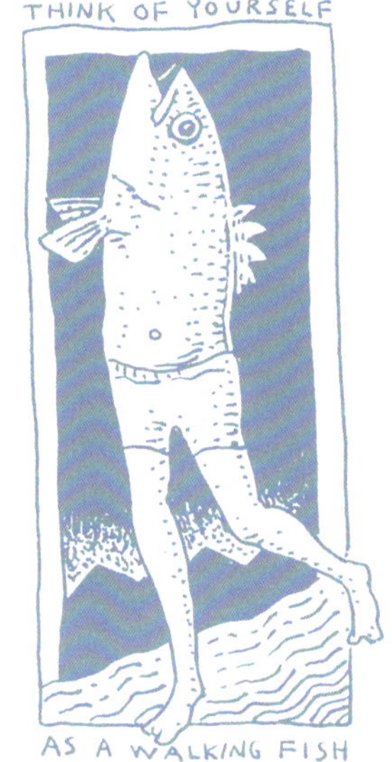

Design and composition Nola Burger **Text** 10/13 Filosofia **Display** Tasse Bold **Printer and binder** Friesens Corporation